T0092386

Interdisciplinary Applied Mathematics

Volume 58

Problems in engineering, computational science, and the physical and biological sciences are using increasingly sophisticated mathematical techniques. Thus, the bridge between the mathematical sciences and other disciplines is heavily traveled. The correspondingly increased dialog between the disciplines has led to the establishment of the series: Interdisciplinary Applied Mathematics.

The purpose of this series is to meet the current and future needs for the interaction between various science and technology areas on the one hand and mathematics on the other. This is done, firstly, by encouraging the ways that mathematics may be applied in traditional areas, as well as point towards new and innovative areas of applications; and secondly, by encouraging other scientific disciplines to engage in a dialog with mathematicians outlining their problems to both access new methods as well as to suggest innovative developments within mathematics itself.

The series will consist of monographs and high-level texts from researchers working on the interplay between mathematics and other fields of science and technology.

Stephen S.-T. Yau • Xin Zhao • Kun Tian •
Hongyu Yu

Mathematical Principles in Bioinformatics

 Springer

Stephen S.-T. Yau
Department of Mathematical Sciences
Tsinghua University
Beijing, China

Yanqi Lake Beijing Institute of
Mathematical Sciences and Applications
(BIMSA)
Beijing, China

Kun Tian
Department of Mathematics
Renmin University of China
Beijing, China

Xin Zhao
Beijing Electronic Science and Technology
Institute
Beijing, China

Hongyu Yu
Department of Mathematical Sciences
Tsinghua University
Beijing, China

ISSN 0939-6047   ISSN 2196-9973   (electronic)
Interdisciplinary Applied Mathematics
ISBN 978-3-031-48294-6   ISBN 978-3-031-48295-3   (eBook)
https://doi.org/10.1007/978-3-031-48295-3

This Springer imprint is published by the registered company Springer Nature Switzerland AG
The registered company address is: Gewerbestrasse 11, 6330 Cham, Switzerland

Paper in this product is recyclable.

Preface

Bioinformatics is among the newest and most highly demanded subjects in biology and mathematics. In recent years, however, bioinformatics has undergone vast changes in style and language. For these reasons, the subject has gained a reputation for inaccessibility. This book presents the main laws of general biology in a rigorous mathematical setting, accompanied by, and indeed with special emphasis on, applications to the study of various interesting biological problems and the development of useful computational tools. I hope that this book will serve two purposes. The first one is to introduce the subject of bioinformatics to readers without a biology background. The second one is to give biologists valid mathematical tools for biological analysis and to help biologists understand how to use these mathematical tools to deal with bioinformatics problems.

Many studies in molecular biology require performing specific computational procedures on given sequence datasets. Traditional textbooks and bioinformatics methods are normally based on multiple sequence alignment (MSA) analysis. However, there are two fundamental drawbacks to alignment methods. The first one is that these methods are based on the choice from a variety of models. How the model is chosen is somewhat arbitrary, and if the wrong model is chosen, the results may be inaccurate. Second, using the language of computational complexity theory, the MSA algorithm is non-deterministic polynomial-time hard (NP-hard). As a result, it is unable to cluster a dataset consisting of thousands of sequences, not to mention the million sequence datasets required in modern molecular biology.

When I started teaching bioinformatics courses, I quickly realized that none of the textbooks that I found on the subject covered the above problems. Those textbooks were not very satisfying from a mathematician's point of view and were unacceptable for my purposes. I was looking for fundamental laws which govern seemingly complicated biological phenomena. What I needed was a clear and mathematically rigorous exposition of the procedures, algorithms, and principles commonly used in bioinformatics. Consequently, I began writing my own lecture notes. It is these notes, together with my research work, which have been published in various journals, that form the basis of this book. I used these materials to teach bioinformatics courses at the Department of Mathematics, Statistics, and Computer

Science of the University of Illinois at Chicago for two years before 2011. I also used these materials to teach at the Department of Mathematical Sciences of Tsinghua University in Beijing, China in 2013, 2018, 2019, and 2022.

This book can be divided into two parts. The first part has three chapters that provide an introduction to the subject of bioinformatics for readers with good mathematical backgrounds but insufficient biological backgrounds. This first part covers basic background knowledge in molecular biology, some useful bioinformatics databases, and the concept of sequence alignment. Chapter 1 covers nucleotides, DNA, RNA, proteins, the genetic code, and how DNA translates into proteins. Chapter 2 describes commonly used databases, such as nucleotide sequence databases, protein sequence databases, and sequence motif databases. Chapter 3 is concerned with alignment methods, for example, global alignment, local alignment, and multiple alignments.

The second part of the book consists of five chapters that describe several bioinformatics principles using a rigorous mathematical formulation. The aim of this part is to give biologists valid mathematical tools for understanding biology. Chapter 4 introduces the time-frequency spectral principle and its applications in bioinformatics. In Chaps. 5 and 6, we use two-dimensional graphical representations which allow people to visually compare different DNA sequences or different protein sequences and see their differences. We also introduce higher dimensional natural vectors which are able to represent DNA and protein sequences without losing any information. This provides real-time, accurate complete sequence comparison and phylogenetic analysis. Chapter 7 presents the convex hull principle and shows how it can be used to mathematically determine whether a certain amino acid sequence can be a protein. The last chapter summarizes additional mathematical ideas relating to sequence comparisons, such as new feature vectors and metrics.

Our book discusses and gives principles that relate to the current open problems in bioinformatics. One guiding light is the convex hull principle. It states that the convex hulls formed from the natural vectors of the genomes/proteins from the same family do not intersect with convex hulls of natural vectors from other families. This principle implies that biological sequences with similar distributions of the nucleotides or amino acids should be in the same family. This opens up a new interdisciplinary area of research in biology, mathematics, and computer science. We believe that our convex hull principle will become a powerful tool in protein research and it can be viewed as a biological law. Another important principle is how to determine whether an arbitrary amino acid sequence can be a protein sequence. No one has developed a criterion for this before. We show that when the collection of arbitrary amino acid sequences is viewed in an appropriate geometric context, the protein sequences cluster together. This leads to a new computational test that has proved to be remarkably accurate at determining whether an arbitrary amino acid sequence can be a protein. We believe our computational test will be useful for researchers who are attempting to complete the job of discovering all proteins or constructing the protein universe. The 3-base periodicity principle is a useful tool that is also introduced in our book. The 3-base periodicity, identified as a pronounced peak at the frequency $N = 3$ of the Fourier power spectrum of the DNA sequences, is prevalent in most exon sequences, but not in intron sequences. This gives a new understanding of gene prediction study based on Fourier spectral analysis. We also introduce effective ways of performing sequence comparisons and phylogenetic analyses. DNA and protein sequences can be represented geometrically as two-dimensional graphs or algebraically as high-dimensional natural vectors without losing any information about the sequence. The two representations are in one-one correspondence, and you can convert between a sequence and either of its representations without losing information. The similarity between two sequences can be calculated using either representation. The Yau-Hausdorff distance is used for calculating similarity for the graph representation, and the usual Euclidean distance is used for the natural vector representation similarity calculation. This creates a new space with biological distance, which allows us to do phylogenetic analysis in a most natural and efficient manner. To be self-contained, the book includes a description of the Yau-Hausdorff distance.

Among the above principles, the convex hull principle is of the greatest significance and provides a solution to one of the 23 Mathematical Challenges of DARPA (The Defense Advanced Research Projects Agency): "What are the Fundamental Laws of Biology?" This principle, along with the outstanding performance of the natural vectors in the classification problem, shows that the geometry of the space formed by natural vectors can reflect the relationship among biological sequences well. It also solves another problem of DARPA: "The Geometry of Genome Space."

Overall, the book concentrates on mathematical methods in bioinformatics including plenty of natural ideas and my own research progress in the area. There is no available book that discusses the fundamental principle of bioinformatics. Most of the books in the market rely on some artificial assumptions, the so-called model approach. Ours is the first book that develops the fundamental principles of bioinformatics. Although this book is primarily a textbook for students with some mathematical background, at the same time, it is also suitable for any mathematician or biologist, or anyone who is interested in mathematical principles in bioinformatics. The mathematical knowledge used in the book is explained in detail.

We would like to thank my friend Prof. Alexander Isaev of the Department of Mathematics of the Australian National University (ANU) in Canberra for his encouragement and many valuable comments. In fact, the Chap. 3 Sequence Alignment was written with his generous help. I also want to thank all my previous Ph.D. students in the subject of Bioinformatics. Much of the work described in Chaps. 4 through 8 was done by them under my guidance.

My father passed away when I was 11 years old. I am very grateful to my mother, Yau-Leung Yeuk Lam, who was determined to give me the chance to receive a high-level education. Without her continuous support, I would not be able to make this contribution to the field of bioinformatics today. I owe so much to her for who I am and who I shall be.

Beijing, China   Stephen S.-T. Yau
April, 2023

Acknowledgments

Stephen S.-T. Yau is supported by National Natural Science Foundation of China (NSFC) grant (12171275) and Tsinghua University Education Foundation fund (042202008). Xin Zhao is supported by National Natural Science Foundation of China (NSFC) grant (12201015).

Contents

Chapter 1
Biological Overview

The purpose of this chapter is to provide some basic background in molecular biology for readers. Among the molecules in cells, macromolecules, which are large molecules made by joining small molecules (monomers) into polymers, are of most interest to us. There are three types of macromolecules: DNA, RNA, and protein. In this chapter, we will introduce their components, structures, and properties and show how DNA templates RNA and protein [1].

1.1 Basic Information on Macromolecules

In this section, we only introduce some basic concepts of three types of macro-molecules. Further information will be shown in later sections.

Nucleic acids are the polymers of nucleotides. They can be divided into two classes: deoxyribonucleic acids (DNA) and ribonucleic acids (RNA). Both of them have a distinguishable direction from one end called $5'$ to the other end called $3'$. (The reason why they are named $5'$ and $3'$ will be provided in Sect. 1.4.)

DNA is the basis of heredity and all other macromolecules are made from the instruction of DNA, directly or indirectly. Nucleobases of nucleotides in DNA include adenine (A), cytosine (C), guanine (G), and thymine (T). In most cases, DNA exists as a double-stranded form. Two strands are connected by hydrogen bonds between complementary bases in each strand (A matches T and C matches G). Two strands form a helix structure and the direction of the two strands are opposite. (See Fig. 1.1.) Therefore, the relationship between the two strands is a reverse complement.

The nucleotides of RNA are different from those of DNA in two aspects. First, the nucleobases thymine (T) is replaced by uracil (U). In other words, RNA is written as a string of letters from {A, C, G, U}. Second, the sugar in the nucleotides is

© The Author(s), under exclusive license to Springer Nature Switzerland AG 2023
S. S.-T. Yau et al., *Mathematical Principles in Bioinformatics*, Interdisciplinary Applied Mathematics 58, https://doi.org/10.1007/978-3-031-48295-3_1

Fig. 1.1 Double-stranded
DNA

$$5'\ \text{A T G A T A}\ 3'$$
$$3'\ \text{T A C T A T}\ 5'$$

Table 1.1 Amino acids and their abbreviations

Amino acid	3 letter code	1 letter code	Amino acid	3 letter code	1 letter code	Amino acid	3 letter code	1 letter code
Alanine	Ala	A	Glycine	Gly	G	Proline	Pro	P
Arginine	Arg	R	Histine	His	H	Serine	Ser	S
Aspartic acid	Asp	D	Isoleucine	Ile	I	Threonine	Thr	T
Asparagine	Asn	N	Leucine	Leu	L	Tryptophan	Trp	W
Cysteine	Cys	C	Lysine	Lys	K	Tyrosine	Tyr	Y
Glutamic acid	Glu	E	Methionine	Met	M	Valine	Val	V
Glutamine	Gln	Q	Phenylalanine	Phe	F			

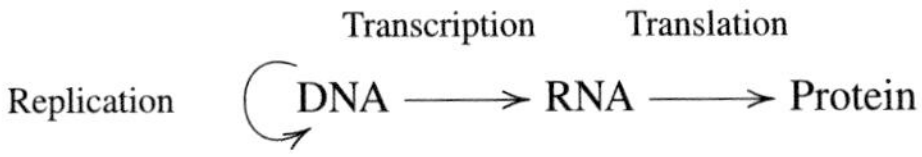

Fig. 1.2 The central dogma

Replication $\big($ DNA $\xrightarrow{\text{Transcription}}$ RNA $\xrightarrow{\text{Translation}}$ Protein

different. In addition, RNA usually appears in the form of a single strand while DNA is double-stranded.

Protein is a kind of polymer made of 20 types of amino acids with directionality. Table 1.1 is the list of the amino acids and their abbreviations.

1.2 The Central Dogma

In 1958, Francis Crick put forward the "central dogma" to summarize the information flow in a cell. The information here refers to the determination of the sequence, either base in the nucleic acid or amino acids' residues in the protein. The direction of information flow shows how one macromolecule is synthesized with the instruction of another macromolecule. Except for some special viruses, the "central dogma" includes three processes: the replication process of DNA, the transcription process to produce RNA, and the translation process to synthesize protein (Fig. 1.2).

Replication is illustrated by the loop from DNA to DNA in the picture. It is the process that DNA duplicates with the guidance of itself in cell division. In replication, two strands in the double helix are separated with a special class of enzymes called helicases and then each strand is used to template a complementary strand from $5'$ to $3'$. In this way two almost identical molecules are made, each having one strand of the original molecule. We say they are almost identical because there are replication errors. By a series of correction mechanisms, the final replication error rate is less than 10^{-9}. The accumulation of replication errors can lead to dysfunction or even cancer for organisms while it is a significant part of evolution (Fig. 1.3).

Fig. 1.3 Replication

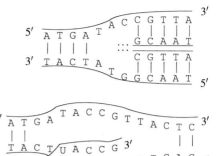

Fig. 1.4 Transcription

Transcription is the process to let information flow from DNA to RNA. In transcription, after separating two strands of the double helix, one strand of DNA will be used to template a single strand of RNA. After this process, the double-stranded DNA remains as before and a single-stranded RNA is synthesized with the instruction of DNA (Fig. 1.4).

Translation is the process that protein is encoded by messenger RNA (mRNA). Specifically speaking, in ribosomes, a type of organelle, mRNA acts as a template and transfer RNA (tRNA) transports amino acids, which are the raw material of protein. They work together to synthesize long polypeptide chains of amino acids and these polypeptide chains will become proteins after folding into particular three-dimensional structures. In translation, three nucleotides (known as a codon) correspond to one amino acid.

From the "central dogma," we can say that DNA is the genetic material of all cells and most viruses. First, organisms use DNA to transfer genetic information to their descendants. Second, other macromolecules (RNA and protein) are all produced with the instruction of DNA, directly or indirectly so DNA contains all the heredity information of the organisms.

The "central dogma" also states that the information flow from DNA to protein is irreversible. It is possible to transfer the heredity information from nucleic acids to nucleic acids or from nucleic acids to proteins, but it is impossible to transfer the heredity information from proteins to proteins, or from proteins to nucleic acids.

Today, the "central dogma" has been extended. Two other information flow paths, RNA replication and reverse transcription, have been found in viruses. Some viruses use RNA as templates to produce new RNA directly and some viruses use RNA to synthesize the complementary DNA (cDNA) first and then use cDNA to produce RNA that can encode protein. The genetic materials of these two types of viruses are RNA.

Fig. 1.5 The structure of the
nucleotide

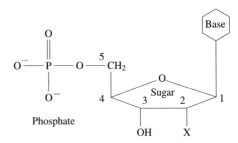

1.3   Nucleotides and Amino Acids

Nucleotides are the monomers that compose DNA and RNA. A nucleotide is composed of three parts: a nitrogen-containing base (nucleobase), a sugar molecule, and a phosphate group. Specifically, the nucleobase connects the $1'$ carbon of the sugar while the phosphate group is bound to the $5'$ carbon. For deoxyribonucleotides, the sugar molecule is deoxyribose and the nucleobase is chosen from adenine (A), cytosine (C), guanine (G), and thymine (T). For ribonucleotides, the sugar molecule is ribose and the nucleobase uracil (U) replaces thymine (T). (See Fig. 1.5. X is H for deoxyribonucleotide, the nucleotides that form DNA, and is OH for ribonucleotide, the nucleotides that form RNA. In the graph, some H molecules are omitted.)

Among five bases, adenine and guanine have a 2-ring structure and are called purines. Cytosine, thymine, and uracil have a 1-ring structure and are called pyrimidines. (See Fig. 1.6.) An important property of nucleobases is that they can form hydrogen bonds between complementary bases (A with T or U and C with G) when forming the double-strand structure in nucleic acids. There are two hydrogen bonds in A-T pairing and A-U pairing and there are three hydrogen bonds in C-G pairing.

There can be more than one phosphate group attached to a nucleotide. Taking off additional phosphate groups produces energy while attaching new phosphate groups absorbs energy. This property is practical for cells to transport energy. In the process of extending DNA or RNA, the raw materials are nucleotides that have three extra phosphate groups (triphosphate nucleotides) since they can provide the energy for extensions.

The chemical structure of the amino acid can be regarded as a central carbon (C_α) connecting an amino group (-NH_2), a carboxyl group (-COOH), and an R group which determines the type of the amino acid. (See Fig. 1.7.) The R group can be also called the residue. There are 20 kinds of residues.

These different residues lead to diverse chemical properties of amino acids. It is believed that one of the most important driving forces in protein folding is the hydrophobic force, which represents the tendency for a molecule to avoid contact with water molecules. Besides the properties of amino acids themselves, the interaction between amino acids such as salt bridges and disulfide bonds is also of great significance.

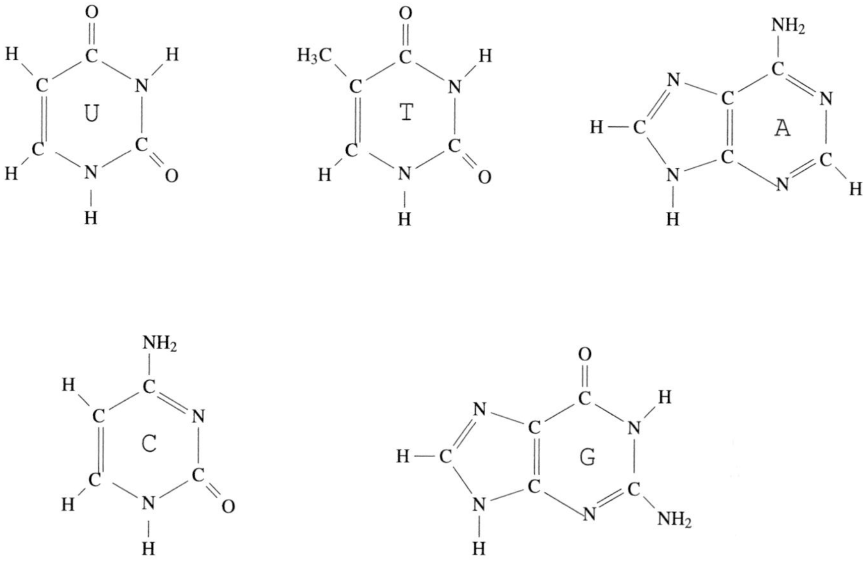

Fig. 1.6 The structure of five nucleobases

Fig. 1.7 The structure of the amino acid

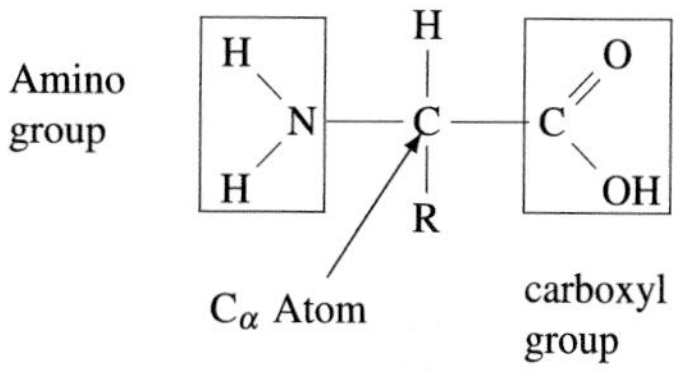

1.4 DNA

Deoxyribonucleotides form a strand via the phosphodiester bond where the phosphate group of one nucleotide is attached to the $3'$ carbon of the deoxyribose of the next nucleotide. There is a direction for each strand. The end where nucleotide has a free phosphate group is called the $5'$ end and the other end is called the $3'$ end. (The number is the carbon atom position where the next nucleotide can be attached.) In nature, the energy for the DNA extension is given by the new nucleotide containing three phosphate groups. Therefore, DNA always extends from $5'$ to $3'$. If the extending direction is opposite, then we need to add phosphate groups to the $5'$ end of the existing strand to provide energy. In this way, removing the first nucleotides on the $5'$ end will be hard because the energy provider will be removed.

In the eukaryotic cell nucleus, DNA is combined with protein to form chromosomes. To be specific, DNA tightly entangles a type of highly conserved protein, histones, in the structure nucleosome and makes chromosomes visible after using certain stains. For each nucleosome, the DNA molecule of about 150–200 bps forms about two superhelices. For high-level organisms, most DNA appears in pairs of

Table 1.2 Notations for nucleotides (including uncertain cases)

Code	Meaning	Complement	Code	Meaning	Complement
A	A	T	Y	C or T	R
C	C	G	K	G or T	M
G	G	C	B	C or G or T	V
T	T	A	D	A or G or T	H
M	A or C	K	H	A or C or T	D
R	A or G	Y	V	A or C or G	B
S	C or G	S	X/N	A or C or G or T	X
W	A or T	W	•	not A, C, G, T	•

linear chromosomes and for many viruses and bacteria, DNA can exist in a circular form.

In sequencing, due to laboratory uncertainties, we sometimes cannot completely determine all bases. For example, we may know a certain base is either A or C but we are not sure which is true. Therefore, we will use some notations to denote these situations. (See Table 1.2.)

The entire set of DNA in a cell can be called a genome. For human beings, there are about 3×10^9 letters in the genome. A natural idea is that there is a strong relationship between the length of the genome and the complexity of organisms. However, it is not always correct. For instance, the lungfish genome is almost 50 times as large as the human genome.

Genes are the basic unit of heredity. A gene can be viewed as the entire sequence of nucleotides that is needed for producing a polypeptide chain or functional RNA. A DNA molecule can contain many genes. It is worth noticing that there exist some parts of a DNA molecule that do not belong to any gene. Another essential point to note is that not all segments within a gene directly encode proteins. These segments are referred to as non-coding regions. Conversely, the segments responsible for coding proteins are known as coding regions.

Biologists have found a counterintuitive fact that the percentage of nucleotides in non-coding regions can be very high for many organisms. Among 23 pairs of chromosomes amounting to about 3×10^9 base pairs in the human genome, there are only about 10^5 genes [2]. We have mentioned that three bases correspond to one amino acid in translation. Assuming that a protein consists of at most 1000 amino acids, then we can infer that 90% of DNA is non-coding [2]. Being non-coding does not mean being useless. Non-coding regions such as promoters and terminators are of great significance for cells.

The promoter is a section of DNA upstream of a gene that RNA polymerases bind when producing RNA. It is the starting point of transcription. The terminator, similarly, is a downstream DNA section that terminates the transcription process. For many prokaryotes, there are two important parts in promoters, named the Pribnow box and the TATA box (Goldberg-Hogness box). We use +1 to denote the transcription start site and −1 to the neighboring site upstream. (There is no 0 in

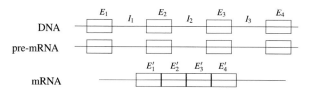

Fig. 1.8 Exons and introns

this scheme, -1 is followed by 1.) The Pribnow box and the TATA box are highly conserved segments, whose centers are located at -10 and -35 approximately, respectively. (The consensus sequences are TATAAT and TTGACA, respectively.) Two boxes are important for RNA polymerases' binding and the exact distance of them influences the activity of genes.

For organisms of low complexity like viruses, the percentage of nucleotides in coding regions is much higher than the human genome. For instance, in the polyomavirus genome, more than 90% of the genome is coding regions [3].

The genes of prokaryotes and eukaryotes are quite different. Cells of prokaryotes do not have a true nucleus or membranous organelles. Cells of eukaryotes, on the other hand, do have a true nucleus to separate DNA from the cytoplasm (the content of cells including organelles) and have membranous organelles. For genes of prokaryotes, the coding regions of DNA are continuous. For genes of eukaryotes, the coding DNA is interrupted by parts that somehow disappeared in the mRNA. As is shown in Fig. 1.8, the RNA segments corresponding to I_1, I_2, and I_3 are removed to get mRNA that produces protein. (The RNA segment E_i' corresponds to E_i for each i.) I_1, I_2, and I_3 are called introns. E_1, E_2, E_3, and E_4, whose heredity information finally pass to protein, are denoted as exons. Exons may be sparse in a gene. For example, in a human gene, thyroglobulin, exons of 8500 bps are interrupted by over 40 introns of 100,000 bps. Determining exons and introns is a significant problem in Bioinformatics.

After removals of introns, sometimes it may still be insertions or removals of nucleotides directed by guide RNA (gRNA). This process is RNA editing. RNA editing has a strong impact on cells since a few changes in the sequence of nucleotides might lead to huge changes for corresponding proteins.

Before polypeptide chains translated by mRNA fold into proteins, there can be a protein splicing process similar to the genes of eukaryotes. The removed parts in the splicing are called inteins and the remaining parts are called exteins.

1.5   RNA

Nucleotides construct a strand of RNA also via phosphodiester bond and the way of connection is the same as that of DNA. In most cases, RNA is single-stranded in the cell. There are mainly three types of RNA: messenger RNA, transfer RNA, and ribosomal RNA. The process of translation is contributed to the efforts of three kinds of RNA together.

Messenger RNA (mRNA) is a single-stranded RNA produced in transcription that carries heredity information from DNA. It plays an important role in translation. The production of protein uses mRNA as templates. In transcription, the RNA polymerase reads the sequence between promoters and terminators and converts it to RNA. For prokaryotes, the RNA produced is mRNA. For eukaryotes, the RNA produced, named primary transcript mRNA or pre-mRNA, contains sections that correspond to introns and still need further processing. After a series of splicing, pre-mRNA is converted to mRNA.

Similar to most RNA molecules, tRNA is also single-stranded without the complementary strand. However, unlike mRNA, tRNA tends to fold back and form hydrogen bonds with itself. To be more specific, tRNA forms a cloverleaf structure by hydrogen bonds and the cloverleaf folds into an L shape in the three-dimensional space. The length of tRNA is usually between 70 and 90 bps. As the amino acids' transporter in the translation process, tRNA has two important functions. The first is the ability to carry the amino acid. Transfer RNA can link the specific amino acid on its acceptor stem located on its $3'$ side. The second is the ability to identify codons. By a part called anticodon, tRNA can form three base pairs with the target mRNA. Besides acceptor stem and anticodon, there are other structures in tRNA such as D-arm and T-arm, but we will not focus on them in this book.

We have mentioned that ribosomes are important organelles in translation. They are the places where tRNA matches the codons of mRNA to produce polypeptide chains. Ribosomes are constituted by ribosomal RNA and protein. Ribosomal RNA is a kind of enzyme. In translation, amino acids are linked to produce polypeptide chains. This process is catalyzed by ribosomal RNA.

There are also some other RNA in living systems [2]. Small nuclear RNA is found in eukaryotic cells' nuclei and is important in processing RNA. For example, in the process where pre-mRNA is spliced to produce mRNA, snRNA plays a significant role. Guide RNA is also used for mRNA editing. It can do point-wise nucleotide insertions or deletions for RNA.

1.6 Protein

Two amino acids can be connected via a peptide bond, where the carboxyl group of one amino acid binds the amino group of the next one. (See Fig. 1.9.) It makes amino acids form a linear polymer called a polypeptide chain. Two ends of the chain can be named by their free group (N terminus and C terminus).

The sequence of amino acids in the polypeptide chain is the primary structure of the protein. Besides the primary structure, polypeptide chains can fold into diverse three-dimensional structures (or called conformations) since there are some angles in the polypeptide chain that can be changed.

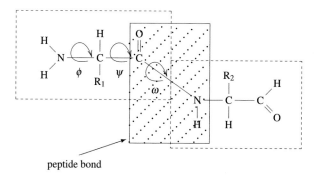

Fig. 1.9 The structure of the peptide bond connected by two amino acids

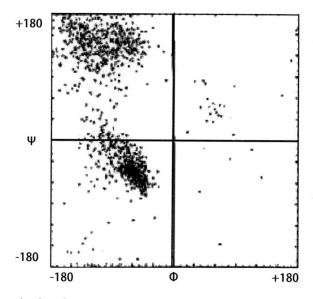

Fig. 1.10 Ramachandran plot

In polypeptide chains, there are three kinds of angles. Since the peptide bond is planar, the angle of the peptide bond, ω, has only two possible choices, which are 180° (trans) and 0° (cis), respectively. Most conformations in polypeptide chains are trans. The other two angles, ϕ and ψ, which are angles of N-C_α bond and C-C_α bond, respectively, have more freedom. Nevertheless, the two angles are not independent. The well-known Ramachandran plot shows that the pair (ϕ, ψ) locates in a restricted area. (See Fig. 1.10. This picture is from [4]. This picture considers 310 proteins and each dot represents a (ϕ, ψ) pair of one amino acid in proteins.)

Given a polypeptide chain, the three-dimensional structure is determined by its order of amino acids in natural cases. This structure is believed to have a global minimum of free energy. (The free energy includes the potential energy of van der Waals attraction, electrostatic forces, hydrogen bonds, and other attractions.)

However, sometimes proteins can be folded incorrectly due to some abnormal influences and these wrongly folded proteins may lead to diseases. Some abnormal proteins can even bind the normal proteins and transform them into a wrong conformation. These infectious proteins are called prions.

To reduce the impact of wrong folding, there are certain proteins called chaperones that can assist in the proper folding of proteins. Chaperones are located in the neighborhood of ribosomes. They can recognize wrongly folded proteins by hydrophobic residues on the protein's surface and help refold abnormal proteins.

The three-dimensional structure of a protein can determine its function. Therefore, two proteins that have completely different sequences but are similar in three-dimensional structure may process similar functions.

Proteins can be divided into three classes roughly by their shapes: globular proteins, fibril proteins, and membrane proteins. Many physiologically active proteins such as enzymes and immunoglobulins are globular proteins. Fibril proteins including collagen and elastin can form strong structures in organisms. Membrane proteins are important for material transport in cell membranes.

In the three-dimensional structure of proteins, some pieces are frequently used such as α helices and β sheets. An α helix is a structure formed by stacking amino acids in a polypeptide chain like a spiral and a β sheet is a linear structure formed by adjacent protein strands and the hydrogen bonds between them. There are also other common pieces such as turns and other helices' structures (π helix, 3_{10} helix). These pieces form the secondary structure of a protein. The combinations of secondary structures form supersecondary structures (also called motifs) such as helix-turn-helix motifs.

The tertiary structure, composed of secondary structures and supersecondary structures, is the structure that has specific functions. At this level, proteins are available to interact with other proteins or other molecules. If more than one protein with a tertiary structure gathers together and forms a protein complex, the quaternary structure appears.

1.7   The Genetic Code

The genetic code is the rule that organisms use to produce specific proteins with the instructions of DNA. In transcription, we know that the heredity information passes from DNA to RNA by the pairing rule of nucleobases. So the genetic code often refers to the code of how protein is encoded by mRNA. It is a question that scientists have studied for a long time since Watson and Crick proposed the double helix model of DNA in 1953.

There are many conjectures about how exactly a sequence of nucleotides encodes a polypeptide chain. We here introduce an interesting attempt from Crick. Crick assumed that the code reads blocks of k letters where $4^k \geq 20$ since there are only four types of nucleobases for RNA and there are 20 types of amino acids. Therefore, the length of the blocks cannot be less than three letters long and Crick took $k = 3$. Crick also believed that each nucleobase should be read in only one block and the reading frame, which is the phase of codon reading, is uniquely determined by the blocks. To be more specific, if $n_1n_2n_3$ is a sequence of nucleotides that encodes amino acid A_1 and $n_4n_5n_6$ is a sequence of nucleotides that encodes amino acid A_2, then $n_2n_3n_4$ and $n_3n_4n_5$ should not encode any amino acid or the meaning of the sequence $n_1n_2n_3n_4n_5n_6$ will be ambiguous. Therefore, in Crick's theory, AAA is impossible to be a codon, or in AAAAAA there will be no obvious reading frame since there are four places to begin reading AAA.

The experiments of biologists show that Crick is partially correct. The genetic code reads three letters as a block to encode one amino acid and the blocks are non-overlapping in one translation process. However, different from Crick's theory, all $4^3 = 64$ possible triplets are codons and there may be different codons that correspond to the same kind of amino acid. This coding method may be less elegant in the view of mathematics but is practical for lives by reducing the impact of mutations. The property that most amino acids correspond to more than one codon is called the degeneracy of codons. Since the reading frame is no longer determined by the sequences in this method, there should be signals for organisms to know where to start and where to stop the translation. There are two special types of codons: the initiation codon and the termination codon. For most organisms, the initiation codon is AUG and for some prokaryotes, GUG and UUG are also the initiation codons. The termination codons are UAG, UAA and UGA. The initiation codons encode amino acids while the termination codons do not. Table 1.3 gives the genetic code in a compact form. (TC is the termination signal and it does not encode any amino acid.) It is worth noticing that the genetic code is not unique. A few organisms have different correspondences between codons and amino acids. What we show in the table is the most common representation.

Table 1.3 The genetic code

1st	2nd				3rd
	U	C	A	G	
U	Phe	Ser	Tyr	Cys	U
	Phe	Ser	Tyr	Cys	C
	Leu	Ser	TC	TC	A
	Leu	Ser	TC	Trp	G
C	Leu	Pro	His	Arg	U
	Leu	Pro	His	Arg	C
	Leu	Pro	Gln	Arg	A
	Leu	Pro	Gln	Arg	G
A	Ile	Thr	Asn	Ser	U
	Ile	Thr	Asn	Ser	C
	Ile	Thr	Lys	Arg	A
	Met	Thr	Lys	Arg	G
G	Val	Ala	Asp	Gly	U
	Val	Ala	Asp	Gly	C
	Val	Ala	Glu	Gly	A
	Val	Ala	Glu	Gly	G

In fact, from the table, we can see that many pairs of codons that differ only in the third position base code for the same amino acid while pairs of codons differing only in the first or second position usually code for different amino acids.

Since each codon has its meaning, there are many possible reading frames for one mRNA. For each DNA strand, there are three reading frames and there are six reading frames together. In other words, for a given sequence of nucleotides, a shift of one letter will lead to a completely different translation result. This phenomenon may happen in reality. Certain viruses can encode two distinct proteins within roughly the same genomic region by a shift in the reading frame.

An interesting fact is that, although the genetic code is universal for most species, different organisms may have different preferences for the usage of certain codons for particular amino acids. Some species even do not have tRNA corresponding to a particular codon.

Finally, we introduce some discussions about the origin of the genetic code. (See [5–8].) A natural idea is that different genetic codes existed in the past and natural selection has promoted the evolution of the genetic code and the translation machinery until they achieved the optimal case. However, F. Crick has proposed a completely different conjecture. Crick believed that there may be a so-called frozen accident after which any further random mutations are lethal for organisms. This theory can explain why the genetic code is almost the same for completely different species.

Chapter 2
Bioinformatics Databases

2.1 Introduction to Bioinformatics Databases

The advent of rapid DNA sequencing technologies has profoundly revolutionized our comprehension of life sciences. The abundance of data generated by these technologies has fueled the expansion of databases that gather and disseminate sequence, structure, and gene expression information. These biological databases amalgamate data from various sources, including scientific experiments, published literature, high-throughput experiments, and computational analyses. They encompass a wide array of research areas, such as genomics (the study of genomes), proteomics (the study of proteins), and metabolomics (the study of chemical processes involving metabolites). Biological databases provide a wealth of information, including gene function, structure, localization (both cellular and chromosomal), clinical effects of mutations as well as similarities of biological sequences and structures. These databases play a pivotal role in aiding scientists to comprehend and elucidate diverse biological phenomena, ranging from the structure and interactions of biomolecules to the complete metabolism of organisms, and even the evolution of species. This knowledge is invaluable in advancing the fight against diseases, facilitating medication development, and unraveling fundamental relationships among species throughout the history of life.

Organizing biological sequences into a database serves another critical purpose: discovering new biology [1]. Useful sequence patterns are preserved over long periods of evolutionary time. When a newly discovered sequence exhibits significant similarity with an existing sequence in the database, there is a high likelihood that their biological functions might also be similar. This comparison process gives rise to new and valuable biological hypotheses, allowing researchers to draw meaningful insights and make important discoveries based on the relationships observed between sequences. By leveraging the power of sequence comparison, scientists can unravel the hidden connections and functionalities within biological

© The Author(s), under exclusive license to Springer Nature Switzerland AG 2023
S. S.-T. Yau et al., *Mathematical Principles in Bioinformatics*, Interdisciplinary
Applied Mathematics 58, https://doi.org/10.1007/978-3-031-48295-3_2

systems, opening up avenues for further exploration and understanding of the complexities of life.

2.2   Nucleotide Sequence Databases

The primary nucleotide sequence databases include:

- European Molecular Biology Laboratory (EMBL) (http://www.ebi.ac.uk/embl/) [9]
- DNA Data Bank of Japan (DDBJ) (http://www.ddbj.nig.ac.jp/) [10]
- GenBank (http://www.ncbi.nlm.nih.gov/) [11]

They constitute the International Nucleotide Sequence Database Collaboration (INSDC, http://www.insdc.org), standing as one of the most celebrated global initiatives in public domain data sharing. Each database group compiles a segment of the overall sequence data contributed from around the world, with all fresh and revised data exchanged daily to ensure harmonization across them [12]. The amount of data in the database is growing very fast. Figure 2.1 illustrates the data growth of EMBL, DDBJ, and GenBank.

2.2.1   EMBL (http://www.ebi.ac.uk/embl/)

EMBL nucleotide sequence database is maintained by the European Bioinformatics Institute (EBI) in Hinxton, Cambridgeshire, UK. It offers an extensive repository of global nucleotide sequencing details, encompassing raw sequencing data, sequence assembly information, and functional annotations. Up to Aug 21, 2022, the EMBL database contains roughly 3.5 billion sequence records.

2.2.2   DDBJ (http://www.ddbj.nig.ac.jp/)

DDBJ Center collects nucleotide sequence data and provides freely available nucleotide sequence data and a supercomputer system, to support research activities in life science. Up to June 2022, the DDBJ database contains around 3.6 billion sequence records.

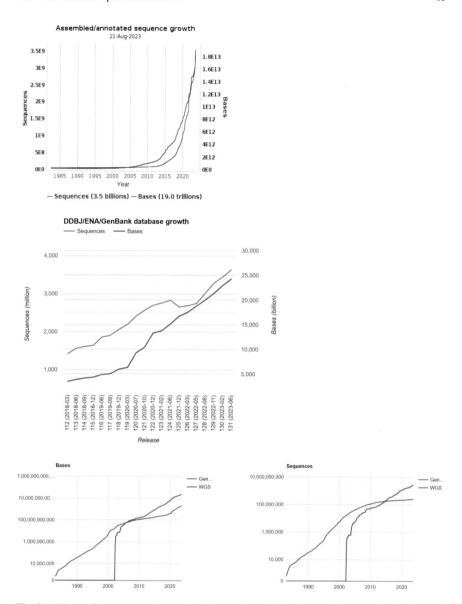

Fig. 2.1 These figures are from https://www.ebi.ac.uk/ena/browser/about/statistics, https://www.ddbj.nig.ac.jp/statistics/index-e.html, and https://www.ncbi.nlm.nih.gov/genbank/statistics/, respectively

2.2.3 GenBank (http://www.ncbi.nlm.nih.gov/genbank/)

The GenBank is the NIH genetic sequence database maintained by the National Center for Biotechnology Information (NCBI, a part of the National Institutes of Health in the United States) [11]. GenBank stands as a curated compilation of all DNA sequences accessible to the public. The indexing of GenBank is organized by release number. Up to Aug 2023, the GenBank database encompasses approximately 246 million sequence records. Starting from 1982 up to the present, the base count within GenBank has doubled roughly every 18 months. Each GenBank entry encompasses a succinct description of the sequence, accompanied by details encompassing the scientific nomenclature and taxonomic classification of the specific organism from which the data is derived. Furthermore, a comprehensive table of features is provided, delineating coding and noncoding regions, alongside other biologically significant sites such as transcription units, points of mutations or modifications, and repetitive elements. Bibliographic references are included for all known sequences with a link to the Medline unique identifier [13]. The following is the list of entries for identifiers:

- **LOCUS:** a short unique name for the entry, chosen to suggest the sequence definition
- **DEFINITION:** a concise description of the sequence
- **ACCESSION:** the primary accession number is a unique, unchanging code assigned to each entry. This code should be used when citing information from GenBank
- **KEYWORDS:** short phrases describing gene products and other information about an entry
- **SEGMENT:** information on the order in which this entry appears in a series of discontinuous sequences from the same molecule
- **SOURCE:** common name of the organism or the name most frequently used in the literature
- **ORGANISM:** formal scientific name of the organism (first line) and taxonomic classification levels (second and subsequent lines)
- **REFERENCE:** citations for all articles containing data reported in this entry. Includes four sub-keywords and may repeat
- **AUTHORS:** lists the authors of the citation
- **JOURNAL:** lists the journal name, volume, year, and page numbers of the citation
- **COMMENT:** cross-references to other sequence entries, comparisons to other collections, notes of changes in LOCUS names, and other remarks
- **FEARURES:** table containing information on portions of the sequence that code for proteins and RNA molecules and information on experimentally determined sites of biological significance
- **BASE COUNT:** summary of the number of occurrences of each base code in the sequence
- **ORIGIN:** specification of how the first base of the reported sequence is operationally located within the genome. Where possible, this includes its location within a larger genetic map

Figure 2.2 gives an example of GeneBank data entry.

Bacillus subtilis strain NDH03 DNA gyrase subunit B (gyrB) gene, partial cds

GenBank: JN590220.1

FASTA   Graphics   PopSet

Go to: ☑

```
LOCUS       JN590220                877 bp    DNA     linear   BCT 11-DEC-2012
DEFINITION  Bacillus subtilis strain NDH03 DNA gyrase subunit B (gyrB) gene,
            partial cds.
ACCESSION   JN590220
VERSION     JN590220.1
KEYWORDS    .
SOURCE      Bacillus subtilis
  ORGANISM  Bacillus subtilis
            Bacteria; Firmicutes; Bacilli; Bacillales; Bacillaceae; Bacillus.
REFERENCE   1  (bases 1 to 877)
  AUTHORS   Guo,Q., Li,S., Lu,X., Li,B., Stummer,B., Dong,W. and Ma,P.
  TITLE     phoR sequences as a phylogenetic marker to differentiate the
            species in the Bacillus subtilis group
  JOURNAL   Can. J. Microbiol. 58 (11), 1295-1305 (2012)
   PUBMED   23145827
REFERENCE   2  (bases 1 to 877)
  AUTHORS   Guo,Q.
  TITLE     PhoP/phoR two-components systems sequences as a phylogenetic marker
            to differentiate the species in genus Bacillus
  JOURNAL   Unpublished
REFERENCE   3  (bases 1 to 877)
  AUTHORS   Guo,Q.
  TITLE     Direct Submission
  JOURNAL   Submitted (17-AUG-2011) Biocontrol for Plant Disease, Plant
            Protection, 437, Dongguan Street, Baoding, Hebei 071000, China
FEATURES             Location/Qualifiers
     source          1..877
                     /organism="Bacillus subtilis"
                     /mol_type="genomic DNA"
                     /strain="NDH03"
                     /db_xref="taxon:1423"
                     /country="USA"
                     /PCR_primers="fwd_name: gyrb-f, fwd_seq:
                     ttgrcgghrgygghtataaagt, rev_name: gyrb-r, rev_seq:
                     tccdccstcagartcwccctc"
     gene            <1..>877
                     /gene="gyrB"
     CDS             <1..>877
                     /gene="gyrB"
                     /codon_start=1
                     /transl_table=11
                     /product="DNA gyrase subunit B"
                     /protein_id="AEW70429.1"
                     /translation="GVGASVVNALSTELDVTVHRDGKIHRQTYKRGVPVTDLEIIGET
                     DHTGTTTHFVPDPEIFSETTEYDYDLLANRVRELAFLTKGVNITIEDKREGQERKNEY
                     HYEGGIKSYVEYLNRSKEVVHEEPIYIEGEKDGITVEVALQYNDSYTSNIYSFTNNIN
                     TYEGGTHEAGFKTGLTRVINDYARKKGLIKENDPNLSGDDVREGLTAIISIKHPDPQF
                     EGQTKTKLGNSEARTITDTLFSTAMETFMLENPDAAKKIVDKGLMAARARMAAKKARE
                     LTRRKSALEISNLPGK"
ORIGIN
        1 ggtgtaggtg cgtcggtcgt aaacgcacta tcaacagagc ttgatgtgac ggttcaccgt
       61 gacggtaaaa ttcaccgcca aacctataaa cgcggagttc cggttacaga ccttgaaatc
      121 attggcgaaa cggatcatac aggaacgacg acacattttg tcccggaccc tgaaattttc
      181 tcagaaacaa ccgagtatga ttacgatctg cttgccaacc gcgtgcgtga attagccttt
      241 ttaacaaagg gcgtaaacat cacgattgaa gataaacgtg aaggacaaga gcgcaaaaat
      301 gaataccatt acgaaggcgg aattaaaagt tatgtagagt atttaaaccg ctctaaagag
      361 gttgtccatg aagagccgat ttacattgaa ggcgaaaagg acggcattac ggttgaagtg
      421 gctttgcaat acaatgacag ctacacaagc aacatttact cgtttacaaa caacattaac
      481 acgtacgaag gcggtaccca tgaagctggc ttcaaaacgg gcctgactcg tgttatcaac
      541 gattacgcca gaaaaaaagg gcttattaaa gaaaatgatc caaacctaag cggagatgac
      601 gtaagggaag ggctgacagc gattatttca atcaaacacc ctgatccgca gtttgagggc
      661 caaacaaaaa caaagctggg caactcagaa gcacggacga tcaccgatac gttattttct
      721 acggcgatgg aaacatttat gctggaaaat ccagatgcag ccaaaaaaat tgtcgataaa
      781 ggtttaatgg cggcaagagc aagaatggct gcgaaaaaag cgcgtgaact aacacgccgt
      841 aagagtgctt tggaaatttc aaacctgccc ggtaagt
//
```

Fig. 2.2 Example of GeneBank data entry: *Bacillus subtilis* strain NDH03 DNA gyrase subunit B (gyrB) gene

2.3   Protein Sequence Databases

UniProt database is the world's most comprehensive repository of protein sequences and functions created by joining the three major protein sequence databases:

- Swiss-Prot (https://www.uniprot.org/uniprotkb?query=reviewed:true) [14]
- Translated EMBL Nucleotide Sequence Data Library (TrEMBL) (https://www. uniprot.org/uniprotkb?query=reviewed:false) [9]
- Protein Information Resource (PIR) (http://pir.georgetown.edu) [15]

It integrates the resources of EBI (European Bioinformatics Institute), SIB (the Swiss Institute of Bioinformatics), and PIR (Protein Information Resource) databases [16]. UniProt comprises three distinct components, each tailored to serve specific functions. The core of UniProt is the UniProt Knowledgebase, which serves as a centralized hub for comprehensive protein-related data encompassing functions, classifications, and cross-references. Additionally, UniProt Reference Clusters (UniRef) amalgamate closely related sequences into unified records, optimizing search efficiency. The UniProt Archive (UniParc) serves as a comprehensive repository that chronicles the evolutionary history of all protein sequences.

2.3.1   Swiss-Prot (https://www.uniprot.org/uniprotkb?query= reviewed:true)

UniProtKB/Swiss-Prot is the manually annotated and reviewed section of the UniProt Knowledgebase [14]. It is a high-quality annotated and non-redundant protein sequence database, which brings together experimental results, computed features, and scientific conclusions. Up to June 2023, the Swiss-Prot database has encompassed approximately 570 thousand sequence entries. The growth of the database is summarized as Fig. 2.3. This database is generally considered one of the best protein sequence databases in terms of the quality of the annotation.

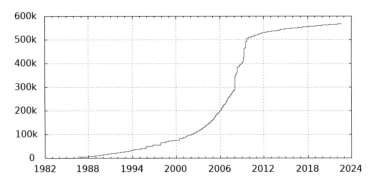

Fig. 2.3 The growth of the database, UniProtKB/Swiss-Prot (https://web.expasy.org/docs/relnotes/relstat.html)

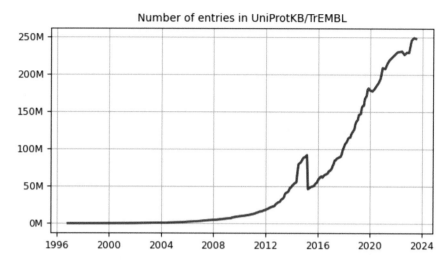

Fig. 2.4 The growth of the database, UniProtKB/TrEMBL (https://www.ebi.ac.uk/uniprot/TrEMBLstats)

2.3.2   TrEMBL (https://www.uniprot.org/uniprotkb?query=reviewed:false)

UniProtKB/TrEMBL (Translated EMBL Nucleotide Sequence Data Library) is an automatically annotated and not reviewed section that contains all the translations of EMBL nucleotide sequence entries not yet integrated into Swiss-Prot. Up to June 2023, TrEMBL database has included about 248 million protein sequences. The growth of the database is summarized in Fig. 2.4.

Swiss-Prot and TrEMBL are developed by the Swiss-Prot groups at EMBL-EBI (European Bioinformatics Institute) and SIB (Swiss Institute of Bioinformatics).

2.3.3   PIR (http://pir.georgetown.edu)

Founded in 1984 by the National Biomedical Research Foundation, the Protein Information Resource (PIR) serves as a valuable tool aiding researchers in identifying and comprehending protein sequence information. It consistently offers cutting-edge resources to facilitate the integration of proteomic and genomic data. Moreover, PIR assumes a pivotal role in advancing the worldwide dissemination and standardization of protein annotation.

2.4 Sequence Motif Databases

The primary sequence motif databases include:

- Pfam (http://pfam.xfam.org/) [17]
- PROSITE (http://prosite.expasy.org/) [18]

In genetics, a sequence motif refers to a prevalent pattern of nucleotides or amino acids with presumed biological significance. For proteins, a sequence motif is distinguished from a structural motif, a motif formed by the three-dimensional arrangement of amino acids, whose corresponding sequences may not be adjacent. Protein sequence motifs serve as distinctive signatures for protein families and are frequently employed as predictive tools for determining protein function.

2.4.1 Pfam (http://pfam.xfam.org/)

Pfam is a database of protein families defined as domains (contiguous segments of entire protein sequences) [17]. For each domain, it contains multiple alignments of a set of defining sequences (the seeds) and the other sequences in Swiss-Prot and TrEMBL that can be matched to that alignment. The database was started in 1996 and is maintained by a consortium of scientists. The Pfam 35.0, released in Nov 2021, encompasses a total of 19 thousand families and clans.

HMMER, a HMM based algorithm developed by Sean Eddy, constitutes the computational framework underpinning Pfam. Pfam offers direct access to the domain architecture of protein sequences from Swiss-Prot and TrEMBL, and users can also explore domain searches for other sequences via web-based servers. The Pfam database can be employed for both searching and domain identification within sequences.

2.4.2 PROSITE (http://prosite.expasy.org/)

PROSITE is a database of protein families and domains [18]. It consists of biologically significant sites, patterns, protein domains, families, and functional sites, as well as associated patterns and profiles to identify them. Its foundation lies in regular expressions that describe characteristic subsequences unique to specific protein families or domains. It is complemented by ProRule, a collection of rules, which increases the discriminatory power of these profiles and patterns by providing additional information about functionally and structurally critical amino acids. Up to June 2023, PROSITE has included a repository of 1311 patterns, 1367 profiles, and 1382 ProRule.

2.5 Macromolecular 3D Structure Databases

The main protein structure databases include:

- Protein Data Bank (PDB) (http://www.rcsb.org/) [19]
- Structural Classification of Proteins (SCOP) (http://scop.mrc-lmb.cam.ac.uk/scop/) [20]
- Class, Architecture, Topology, Homologous superfamily database (CATH) (http://www.cathdb.info) [21, 22]

They offer valuable and profound understandings of molecular function at the atomic scale and furnish conclusive evidence supporting various aspects of molecular function and elucidate patterns of sequence conservation present in protein families. Furthermore, the alignment of 3D structures can be a valuable tool in guiding precise multiple sequence alignments, crucial for conducting phylogenetic analyses [24]. (Phylogenetic analyses involve the examination of the evolutionary history and relationships among organisms or groups of organisms.)

2.5.1 PDB (http://www.rcsb.org/)

The Protein Data Bank (PDB) serves as a main repository for three-dimensional structural data pertaining to significant biological macromolecules like proteins and nucleic acids [19]. This data is usually acquired through methodologies such as X-ray crystallography, NMR spectroscopy, or the emerging technique of cryo-electron microscopy. Up to Aug 2023, the PDB has contained 208 thousand experimental structures.

The PDB entries contain the atomic coordinates, and some structural parameters connected with the atoms (B-factors, occupancies), or computed from the structures (secondary structure). The PDB entries contain some annotation, but it is not as comprehensive as in Swiss-Prot. Fortunately, there are cross-links between the databases in both file formats. Files in the PDB contain various essential information, including the compound's name, the species and tissue it originates from, amino acid sequence, secondary structure locations, and the coordinates of the atoms in the protein. (Hydrogen atoms' coordinates are not included in the PDB due to the constraints of x-ray crystallography and NMR structure analysis.) An example of a PDB file is shown in Fig. 2.5 and its visualization is shown in Fig. 2.6.

The following is the list of entries for identifiers:

- **HEADER, TITLE, and AUTHOR** provide information about the researchers who defined the structure. Numerous other types of records are available to provide other types of information.
- **REMARK** contains free-form annotation, but they also accommodate standardized information. For example, the REMARK 350 BIOMT records describe how to compute the coordinates of the experimentally observed multimer from those of the explicitly specified ones of a single repeating unit.

```
HEADER    OXYGEN TRANSPORT                         22-MAR-79   1MBS
TITLE     X-RAY CRYSTALLOGRAPHIC STUDIES OF SEAL MYOGLOBIN. THE
TITLE     2 MOLECULE AT 2.5 ANGSTROMS RESOLUTION
...
EXPDTA    X-RAY DIFFRACTION
AUTHOR    H.SCOULOUDI
...
REMARK 350 BIOMOLECULE: 1
REMARK 350 AUTHOR DETERMINED BIOLOGICAL UNIT: MONOMERIC
REMARK 350 APPLY THE FOLLOWING TO CHAINS: A
REMARK 350   BIOMT1   1  1.000000  0.000000  0.000000        0.00000
REMARK 350   BIOMT2   1  0.000000  1.000000  0.000000        0.00000
REMARK 350   BIOMT3   1  0.000000  0.000000  1.000000        0.00000
...
SEQRES   1 A  153  GLY LEU SER ASP GLY GLU TRP HIS LEU VAL LEU ASN VAL
SEQRES   2 A  153  TRP GLY LYS VAL GLU THR ASP LEU ALA GLY HIS GLY GLN
SEQRES   3 A  153  GLU VAL LEU ILE ARG LEU PHE LYS SER HIS PRO GLU THR
SEQRES   4 A  153  LEU GLU LYS PHE ASP LYS PHE LYS HIS LEU LYS SER GLU
...
ATOM      1  N   GLY A   1      15.740  11.178 -11.733  1.00  0.00           N
ATOM      2  CA  GLY A   1      15.234  10.462 -10.556  1.00  0.00           C
ATOM      3  C   GLY A   1      16.284   9.483  -9.998  1.00  0.00           C
...
HETATM 1226  CHB HEM A 154      11.541 -10.200   7.336  1.00  0.00           C
HETATM 1227  CHC HEM A 154       9.504  -6.500   9.390  1.00  0.00           C
...
CONECT 1225 1229 1256
CONECT 1226 1232 1239
CONECT 1227 1242 1246
...
MASTER        238    2    1    8    0    0    2    6 1266    1   43   12
END
```

Fig. 2.5 The PDB file of the protein with ID=1MBS

- **SEQRES** gives the sequences of the three peptide chains (named A, B, and C), which are very short in this example but usually span multiple lines.
- **ATOM** describes the coordinates of the atoms that are part of the protein. For example, the first ATOM line above describes the alpha-N atom of the first residue of peptide chain A, which is a proline residue. The first three floating point numbers are its x, y, and z coordinates. The next three columns are the occupancy, temperature factor, and element name, respectively.
- **HETATM** describes coordinates of hetero-atoms, that is those atoms that are not part of the protein molecule. Through the years the file format has undergone many changes and revisions. Its original format is dictated by the width of computer punch cards (80 columns).

2.5.2   SCOP (http://scop.mrc-lmb.cam.ac.uk/scop/)

The Structural Classification of Proteins (SCOP) database is a manually curated classification system for protein structural domains, established to discern structural and amino acid sequence similarities among proteins [20]. This classification

Fig. 2.6 The visualization of
the protein with ID=1MBS

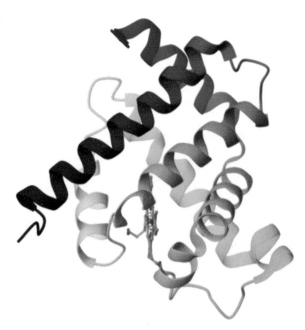

primarily serves the purpose of unraveling evolutionary relationships between proteins. Proteins exhibiting analogous structures yet possessing limited sequence or functional likeness are grouped into distinct "superfamilies," implying a remote common ancestry. Conversely, proteins with comparable shapes and some degree of sequence and/or functional resemblance are categorized into "families," suggesting a closer shared lineage. SCOP's central aim is to organize protein 3D structures hierarchically within structural classes.

Launched by Alexey Murzin in 1994 at the Laboratory of Molecular Biology, MRC, Cambridge, UK, the SCOP database is meticulously curated by experts, ensuring high-quality categorization. Serving as a secondary database, SCOP builds upon primary database (PDB) data through thorough analysis and organization. It employs a hierarchical structure encompassing folds, superfamilies, and families as classifications for protein 3D structures. The latest update of SCOP is in June 2022. In this version, SCOP encompasses 72 thousand distinct domains representing 861 thousand protein structures.

2.5.3   CATH (http://www.cathdb.info)

The CATH (Class, architecture, topology, homologous superfamily) database is a hierarchical classification of protein domain structures, which clusters proteins at four major structural levels [21, 22]. Although the aim is very similar to SCOP, the approach undertaken diverges, and the underlying philosophy when conducting classification exhibits notable distinctions. Notably, a more substantial portion of the determinations involved in classifying novel 3D protein structures is automated by the software.

CATH has employed a crucial algorithm named CATHEDRAL, which serves to ascertain domain boundaries automatically and consequently enhance the frequency of CATH updates. The foundational concept of CATHEDRAL is to identify recurring folds that are already classified within the CATH database.

The CATH database presently encompasses a vast compilation of 151 million protein domains, meticulously categorized into 5841 superfamilies. As an illustrative example, Fig. 2.7 showcases the CATH classification of the protein bovine papillomavirus-1 E2 DNA-binding domain.

CATH Classification

Level	CATH Code	Description
⊚	3	Alpha Beta
Ⓐ	3.30	2-Layer Sandwich
ⓣ	3.30.70	Alpha-Beta Plaits
ⓗ	3.30.70.330	RRM (RNA recognition motif) domain

CATH Clusters

Superfamily	3.30.70.330
Functional Family	Regulatory protein E2

Fig. 2.7 CATH classification of protein bovine papillomavirus-1 E2 DNA-binding domain (http://www.cathdb.info/version/latest/domain/2bopA00). PDB ID = 2BOP, CATH domain ID = 2bopA00, UniProtId = P03122

2.5.4   DALI (http://ekhidna2.biocenter.helsinki.fi/dali/)

In our previous discussion, we introduced three protein structure databases. Now, our focus shifts to the server DALI that help search the similar structures. To be more specific, if you are interested in identifying the structural neighbors of a protein that is already present in the Protein Data Bank, you can access this information from the Dali server [23]. It is based on an all-against-all 3D structure comparison of protein structures in the Protein Data Bank. The structural neighborhoods and alignments are automatically maintained and regularly updated using the Dali search engine [25].

Chapter 3
Sequence Alignment

3.1 Sequence Similarity

New biological sequences do not emerge de novo in nature but rather are derived from pre-existing sequences. This foundational principle underlies sequence analysis. When we establish a connection between a newly discovered sequence and one for which certain information (such as structure or function) is available, we open the possibility of applying the known information, to some extent, to the new sequence as well. Sequences that share a common ancestral origin during evolution are considered related and are termed homologous. It is important to note that, in the context of a belief that all life on Earth originates from a common source, all sequences are fundamentally homologous. However, in practical terms, two sequences are classified as homologous if their relatedness can be verified using some specific methodologies. Therefore, the concept of sequence homology is dynamic, and the delineation of known families of homologous sequences may evolve with improvements in methodologies [26].

The approach for establishing homology can be seen as the method for defining sequence similarity. Different approaches to defining similarity lead to varying outcomes in terms of homology. Among these, sequence alignment stands out as the most widely used method for establishing similarity.

As sequences undergo evolution, their individual residues can undergo three primary types of changes: substitutions, insertions, and deletions. Substitutions involve replacing one residue with another. Insertions involve adding a new residue to the sequence, and deletions involve the removal of an existing residue. Although other evolutionary events, such as segment duplications, inversions, and translocations, do occur, these events are on a larger scale and are less frequent. Typically, these events are not taken into consideration in most sequence alignments.

When we only consider substitutions, the alignment process becomes relatively straightforward. We merely compare two sequences of equal length, checking whether corresponding elements are identical. However, the inclusion of insertions

© The Author(s), under exclusive license to Springer Nature Switzerland AG 2023
S. S.-T. Yau et al., *Mathematical Principles in Bioinformatics*, Interdisciplinary
Applied Mathematics 58, https://doi.org/10.1007/978-3-031-48295-3_3

and deletions necessitates introducing gaps in the alignment, making the alignment calculations more intricate.

For illustration purposes, let us consider the following two nucleotide sequences, which consist of only seven residues each:

$$x : G\ C\ T\ T\ C\ A\ G$$
$$y : T\ T\ T\ A\ G\ C\ C.$$

The sequences have equal lengths, so there exists just one way to align them if no gaps are allowed. That is, we only need to count that there are only one identical element T.

However, if gaps are allowed, there are plenty of possible alignments. For instance, the following alignment appears to be much more acceptable than the preceding one:

$$x : G\ C\ T\ T\ C\ A\ G\ -\ -$$
$$y : T\ -\ T\ T\ -\ A\ G\ C\ C.$$

This alignment suggests that the subsequence $TTAG$ may be an evolutionarily conserved region, which means that both x and y may have evolved from a common ancestral sequence containing this subsequence. There are other reasonable alignments such as:

$$x : G\ C\ T\ T\ C\ A\ G\ -\ -$$
$$y : -\ -\ T\ T\ T\ A\ G\ C\ C.$$

How does one decide among all possible alignments? Addressing these queries necessitates the ability to assign scores to each alignment. The alignments with the highest scores are, by definition, the optimal choices. (Note that there could be multiple optimal alignments.)

In elementary scoring schemes, columns within an alignment are assumed to be independent. As a result, the total score is the sum of column scores. Such schemes require specification of scores $s(a, b) = s(b, a)$ and the gap penalty $s(-, a) = s(a, -)$, with $a, b \in \mathcal{Q}$, where $\mathcal{Q}$ signifies the DNA or RNA alphabet comprising four letters, or the amino acid alphabet with twenty letters. It is noteworthy that the optimal alignments for a sequence pair hinge upon the chosen scoring scheme. Consequently, two different scoring schemes may yield notably distinct optimal alignments. A possible scoring scheme example involves setting $s(a, a) = 1$ for a match, $s(a, b) = -1$ when $a \neq b$ for a mismatch, and $s(-, a) = s(a, -) = -2$ as the gap penalty.

The values $s(a, b)$ form a matrix. In the case of protein alignments, the relationship between two amino acids in substitutions is more intricate. To elaborate, certain substitutions can occur more frequently than others. Thus, relying solely on an identity matrix, where $s(a, a) = 1$ and $s(a, b) = -1$ when $a \neq b$, might yield imprecise results. Previous research has introduced effective substitution matrices like PAM [27] and BLOSUM [28] to address this. We do not delve into the derivation of these matrices since the selection of these matrices does not impact the methodology of sequence alignments.

For the purposes of this chapter we fix a substitution matrix and restrict our considerations to DNA sequences, thus assuming that $\mathcal{Q} = \{A, C, G, T\}$. This setup will be sufficient to demonstrate the main sequence alignment principles.

3.2 Global Alignment

In this section we consider a linear gap model (i.e., set $s(-, a) = s(a, -) = -d$ for $a \in \mathcal{Q}$, with $d > 0$, so that the score of a gap region of length L is equal to $-dL$) and describe an algorithm, the Needleman-Wunsch algorithm [29], which always finds every optimal global alignment. (Note there are often more than one such alignment.) The idea of the algorithm is called dynamic programming, which denotes the act of simplifying a complex problem by decomposing it into more manageable sub-problems through a recursive approach.

Suppose we have two sequences $x = x_1 x_2 \ldots x_i \ldots x_n$ and $y = y_1 y_2 \ldots y_j \ldots y_m$. Let us construct an $(n + 1) \times (m + 1)$-matrix, which we call F. Its (i, j)th element $F(i, j)$ for $i = 1, \ldots, n$, $j = 1, \ldots, m$ is the score of any optimal alignment between the subsequences $x_1 \ldots x_i$ and $y_1 \ldots y_j$. The element $F(i, 0)$ for $i = 1, \ldots, n$ is the score of aligning the subsequence $x_1 \ldots x_i$ to a gap region of length i. Similarly, the element $F(0, j)$ for $j = 1, \ldots, m$ is the score of aligning the subsequence $y_1 \ldots y_j$ to a gap region of length j. We build F recursively starting with the initial condition $F(0, 0) = 0$ and proceeding to fill the matrix from the top left corner to the bottom right one. Here $F(i, j)$ is calculated as follows:

$$F(i, j) = \max \begin{cases} F(i - 1, j - 1) + s(x_i, y_j), \\ F(i - 1, j) - d, \\ F(i, j - 1) - d. \end{cases}$$

Indeed, there are three possibilities for obtaining the optimal score $F(i, j)$: x_i is aligned to y_j (the first option in the formula above), or x_i is aligned to a gap (the second option), or y_j is aligned to a gap (the third option). When calculating $F(i, j)$ we keep the pointer to each of the options from which $F(i, j)$ was derived. When we reach $F(n, m)$, we trace back the pointers to obtain the optimal alignments. The value $F(n, m)$ is then exactly their score. Notice that more than one pointer may come out of a cell of the matrix, which leads to several optimal alignments.

Example 3.1 Let $x = TGGCAC$, $y = AGCC$, and suppose that we are using the scoring scheme $s(a, a) = 1$, $s(a, b) = -1$, if $a \neq b$, and $s(-, a) = s(a, -) = -2$. The corresponding matrix F is shown in Fig. 3.1.

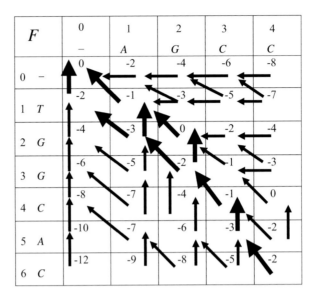

Fig. 3.1 The corresponding matrix F and paths through the matrix of Example 3.1

Tracing back the pointers yields the following three optimal alignments

$$x : T \ G \ G \ C \ A \ C$$
$$y : A - G \ C - C,$$

$$x : T \ G \ G \ C \ A \ C$$
$$y : A \ G - C - C,$$

$$x : T \ G \ G \ C \ A \ C$$
$$y : - A \ G \ C - C$$

whose score is -2. The corresponding paths through the matrix F are shown by thicker arrows.

3.3   Local Alignment

A more biologically relevant alignment problem is the one of finding all pairs of subsequences of two given sequences having the highest-scoring alignments. We will only look at segments, i.e., subsequences of consecutive elements. Any segment of a sequence $x_1 x_2 \ldots x_n$ is of the form $x_i x_{i+1} \ldots x_{i+k}$ where $i + k \leq n$. This alignment problem is usually called the local alignment problem. Here we present the Smith-Waterman algorithm [30], which completely solves the question for a linear gap model.

As in the previous section, we construct an $(n + 1) \times (m + 1)$-matrix, but the formula for its entries is slightly different:

$$F(i, j) = \max \begin{cases} 0, \\ F(i - 1, j - 1) + s(x_i, y_j), \\ F(i - 1, j) - d, \\ F(i, j - 1) - d. \end{cases} \tag{3.1}$$

Opting for the initial choice in the formula mentioned above entails commencing a new alignment. If the optimal alignment achieved until a certain juncture yields a negative score, it becomes more advantageous to initiate a new alignment rather than extending the existing one.

Another distinction from the Needleman-Wunsch algorithm is that in this context, an alignment can terminate at any position within the matrix. Therefore, instead of considering the value $F(n, m)$ located in the bottom right corner of the matrix for the optimal score, we identify the maximum elements within matrix F and initiate the traceback process from these points. The traceback procedure concludes upon reaching a cell with a value of 0, signifying the starting point of the alignment.

Example 3.2 For the sequences from Example 3.1 the only best local alignment is

$$x : G \ C$$
$$y : G \ C,$$

and its score is 2. The corresponding dynamic programming matrix F is shown in Fig. 3.2, where the thicker arrows represent the traceback. Notice that if an element of F is 0 and no arrows come out of the cell containing the element, then it is obtained as the first option in Formula (3.1).

3.4 Alignment with Affine Gap Model

In this section we consider an affine gap model, that is, we let the score of any gap region of length L be equal to $-d - e(L - 1)$ for some $d > 0$ and $e > 0$. In this situation $-d$ is called the gap opening penalty and $-e$ the gap extension penalty. It is common to set e to be smaller than d, which reflects the biological fact that starting a new gap region is harder than extending an existing one. In this section, we only discuss a global alignment algorithm; its local version can be readily obtained as in the preceding section.

The algorithm utilizes three matrices: one $(n + 1) \times (m + 1)$-matrix and two $n \times m$-matrices. For $i = 1, \ldots, n$ and $j = 1, \ldots, m$, let $M(i, j)$ be the score of any optimal alignment between $x_1 \ldots x_i$ and $y_1 \ldots y_j$ under the assumption that the

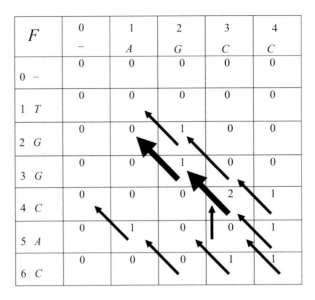

F	0	1	2	3	4
	$-$	A	G	C	C
0 $-$	0	0	0	0	0
1 T	0	0	0	0	0
2 G	0	0	1	0	0
3 G	0	0	1	0	0
4 C	0	0	0	2	1
5 A	0	1	0	0	1
6 C	0	0	0	1	1

Fig. 3.2 The dynamic programming matrix F with the traceback arrows

alignment ends with x_i aligned to y_j. Next, the element $M(i, 0)$ for $i = 1, \ldots, n$ is the score of aligning $x_1 \ldots x_i$ to a gap region of length i and the element $M(0, j)$ for $j = 1, \ldots, m$ is the score of aligning $y_1 \ldots y_j$ to a gap region of length j. Further, for $i = 1, \ldots, n$ and $j = 1, \ldots, m$ define $I_x(i, j)$ to be the score of any optimal alignment between $x_1 \ldots x_i$ and $y_1 \ldots y_j$ provided the alignment ends with x_i aligned to a gap. Finally, for $i = 1, \ldots, n$ and $j = 1, \ldots, m$ set $I_y(i, j)$ to be the score of any optimal alignment between $x_1 \ldots x_i$ and $y_1 \ldots y_j$ provided the alignment ends with y_j aligned to a gap. It follows that if we assume that a deletion is never followed directly by an insertion. In other words, there are no instances where "$a-$" aligns with "$-b$". Otherwise, we can substitute it with the match where a aligns with b, resulting in an increased score.

We can calculate the values in three matrices by the following formulas recursively:

$$M(i, j) = \max \begin{cases} M(i - 1, j - 1) + s(x_i, y_j), \\ I_x(i - 1, j - 1) + s(x_i, y_j), \\ I_y(i - 1, j - 1) + s(x_i, y_j), \end{cases}$$

$$I_x(i, j) = \max \begin{cases} M(i - 1, j) - d, \\ I_x(i - 1, j) - e, \end{cases}$$

$$I_y(i, j) = \max \begin{cases} M(i, j - 1) - d, \\ I_y(i, j - 1) - e. \end{cases}$$

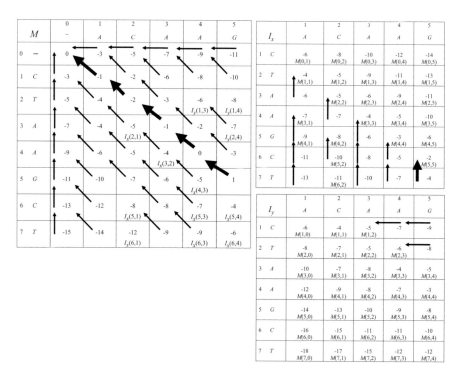

Fig. 3.3 The dynamic programming matrices of Example 3.3

These recurrence relations allow us to fill in the matrices M, I_x, and I_y once we have initialized the process as $M(0,0) = 0$. Here if, for some i and j, one of the options in the right-hand sides of the recurrence relations is not defined (for example, in the formula for $M(1,2)$ the right-hand side contains $I_x(0,1)$ and $I_y(0,1)$), then this option is disregarded in the calculations. The score of an optimal alignment is then equal to $\max\{M(n,m), I_x(n,m), I_y(n,m)\}$, and the traceback starts from the element (or elements) where the maximum is attained.

Example 3.3 Let $x = CTAAGCT$, $y = ACAAG$, the score of any match be equal to 1, that of any mismatch to -1, $d = 3$ and $e = 2$. Then we obtain the dynamic programming matrices displayed in Fig. 3.3.

The arrows and labels show from which elements of the three matrices each number is derived (we also draw the vertical and horizontal arrows in the 0th column and zeroth row of the matrix M). The thicker arrows indicate the traceback; it starts at $I_x(7,5) = -4$. There is only one optimal alignment with score -4:

$$x : C\ T\ A\ A\ G\ C\ T$$
$$y : A\ C\ A\ A\ G\ -\ -.$$

3.5 Heuristic Alignment Algorithms

Previous alignment methods based on dynamic programming are both practical and accurate but come with a drawback of being time-consuming. The time complexity of these algorithms is $O(nm)$ when comparing two sequences of lengths n and m, respectively. (The notation O stands for order. $O(nm)$ signifies that approximately Cnm calculations are required, where C is a constant.) This becomes impractical when dealing with lengthy sequences in computer searches. A similar concern arises for memory complexity. Additionally, when conducting a homology search between a query sequence and an extensive sequence database, these time and memory complexities can pose significant challenges.

To mitigate these challenges, several heuristic algorithms have been developed. These alternatives offer swifter performance but may not necessarily yield the best feasible alignments. In this section, we will delve into some of these heuristic algorithms to address the aforementioned computational constraints.

3.5.1 FASTA

Calculation of dynamic programming matrices takes a substantial amount of time and memory. At the same time, the segments we are interested in are often small in comparison to the entire sequences. The FASTA algorithm is designed to limit the dynamic programming search to particular parts of the matrices [31]. First, FASTA determines the candidate parts, that is, the parts that are likely to lead to optimal alignments. Then, the dynamic programming algorithm is utilized with the additional condition that all matrix elements lying outside these candidate parts are set to $-\infty$. By imposing this condition, the traceback procedure remains confined within the candidate parts. Given that these candidate parts are generally much smaller than the complete matrices, this reduction leads to a substantial enhancement in computational efficiency.

The identification of candidate parts can be achieved through methods such as analyzing the so-called dot matrix. In this matrix, each match between the two sequences is denoted by a dot. Within this matrix, it becomes feasible to pinpoint local similarities by detecting consecutive dots that align diagonally. As illustrated in Fig. 3.4, an instance of a dot matrix is presented, wherein shaded areas represent diagonal runs comprising three or more dots consecutively aligned.

Fix positive integers k and b. Once diagonal stretches of length at least k have been detected, we extend them to complete diagonals, and the candidate parts of the dynamic programming matrix F are then constructed as bands of width b around the extensions. Figure 3.5 shows the candidate parts for the sequences from Fig. 3.4, where $k = 3$ and $b = 1$. (The row and column indexed at 0 are excluded.) The provided illustration follows a linear gap model, while this approach is equally applicable to the affine gap model.

Dot Matrix	A	G	T	C	A	C	T	A	T	G	T	C
T			*				*		*		*	
C				*		*						*
A	*				*			*				
C				*		*						*
A	*				*			*				
G		*								*		
G		*								*		
C				*		*						*
T			*				*		*		*	
A	*				*			*				
G		*								*		
T			*				*		*		*	
C				*		*						*

Fig. 3.4 An example of a dot matrix

F	A	G	T	C	A	C	T	A	T	G	T	C
T					$-\infty$	$-\infty$	$-\infty$	$-\infty$	$-\infty$	$-\infty$	$-\infty$	$-\infty$
C						$-\infty$	$-\infty$	$-\infty$	$-\infty$	$-\infty$	$-\infty$	$-\infty$
A							$-\infty$	$-\infty$	$-\infty$	$-\infty$	$-\infty$	$-\infty$
C								$-\infty$	$-\infty$	$-\infty$	$-\infty$	$-\infty$
A	$-\infty$								$-\infty$	$-\infty$	$-\infty$	$-\infty$
G	$-\infty$	$-\infty$								$-\infty$	$-\infty$	$-\infty$
G	$-\infty$	$-\infty$	$-\infty$								$-\infty$	$-\infty$
C	$-\infty$	$-\infty$	$-\infty$	$-\infty$								$-\infty$
T		$-\infty$	$-\infty$	$-\infty$	$-\infty$							
A			$-\infty$	$-\infty$	$-\infty$	$-\infty$						
G				$-\infty$	$-\infty$	$-\infty$	$-\infty$					
T	$-\infty$				$-\infty$	$-\infty$	$-\infty$	$-\infty$				
C	$-\infty$	$-\infty$				$-\infty$	$-\infty$	$-\infty$	$-\infty$			

Fig. 3.5 The candidate parts for the sequences from Fig. 3.4

It is worth noticing that FASTA searches can miss optimal alignments, with the values of k and b chosen in accordance with the trade-off between time and optimality. A smaller value of k and a larger value of b result in improved performance but require more time, and vice versa.

3.5.2 BLAST

Unlike FASTA, the BLAST (Basic Local Alignment Search Tool) algorithm [32] does not refer to dynamic programming. It looks for short stretches of identities (just as FASTA does) and then tries to extend them in both directions searching for a good longer alignment. This strategy is reasonable from the biological point of view as related sequences tend to share conserved regions. Despite the fact that the search algorithm implemented in BLAST is entirely heuristic, its performance is satisfactory for many cases and is one of the most popular alignment method.

3.6 Multiple Alignment

In sequence analysis it is often required to determine common parts in sequences from a large dataset. To find such a part, one needs to come up with an optimal multiple alignment for the entire dataset. Analogously to the case of a pair of sequences, in order to find an optimal multiple alignment, we need to have a scoring scheme. As in the case of two sequences, the existing alignment methods generally assume that the columns of an alignment not containing gaps are independent and use a scoring function of the form

$$\mathscr{S}(M) = G(M) + \sum_i s(M_i),$$

with M being a multiple alignment, M_i the ith column without gaps, $s(M_i)$ the score of M_i, and G a function for scoring the columns containing gaps.

In the commonly used methods for scoring multiple alignments, the columns not containing gaps are scored by the sum of pairs (SP) function. The SP-score for column M_i not having gaps is defined as

$$s(M_i) = \sum_{k<l} s(M_i^k, M_i^l),$$

where the sum is taken over all pairs (M_i^k, M_i^l), $k < l$, of elements of M_i, and the scores $s(a, b)$, for $a, b \in \mathscr{Q}$, are derived from a substitution matrix used for scoring pairwise sequence alignments. Gaps are frequently scored by letting $s(-, a) = s(a, -)$, setting $s(-, -) := 0$ and defining the corresponding SP-score for the columns containing gaps. We call any such method for scoring gap regions a linear gap model for multiple alignments. Although summing up all the pairwise substitution scores may seem to be natural, in fact there is no statistical basis for an SP-score.

Upon fixing a scheme for scoring multiple alignments, one can generalize pairwise dynamic programming algorithms to aligning n sequences for any $n \geq 3$.

In situations involving multiple alignments one is usually interested in global alignments, and below we give a variant of the Needleman-Wunsch algorithm. Here we assume a scoring scheme with

$$\mathscr{S}(M) = \sum_i s(M_i), \tag{3.2}$$

where the sum taken over all columns (including the ones containing gaps). Note that there is also a multi-dimensional dynamic programming algorithm with the affine gap model.

Suppose one has n sequences $x^1 = x_1^1 \ldots x_{m_1}^1$, $x^2 = x_1^2 \ldots x_{m_2}^2$, ..., $x^n = x_1^n \ldots x_{m_n}^n$. Let $i_1, \ldots, i_n$ be nonnegative integers with $i_j \le m_j$, $j = 1, \ldots, n$, where at least one number is non-zero. Define $F(i_1, \ldots, i_n)$ to be the maximal score of an alignment of the subsequences ending with $x_{i_1}^1 \ldots x_{i_n}^n$ (if for some j one has $i_j = 0$, then the other subsequences are aligned to a gap region). The recursion step of the dynamic programming algorithm is then written as

$$F(i_1, \ldots, i_n) = \max \begin{cases} F(i_1 - 1, \ldots, i_n - 1) + s(x_{i_1}^1, \ldots, x_{i_n}^n), \\ F(i_1, i_2 - 1 \ldots, i_n - 1) + s(-, x_{i_2}^2, \ldots, x_{i_n}^n), \\ F(i_1 - 1, i_2, i_3 - 1 \ldots, i_n - 1) + s(x_{i_1}^1, -, x_{i_3}^3, \ldots, x_{i_n}^n), \\ \vdots \\ F(i_1 - 1, \ldots, i_{n-1} - 1, i_n) + s(x_{i_1}^1, \ldots, x_{i_{n-1}}^{n-1}, -), \\ F(i_1, i_2, i_3 - 1 \ldots, i_n - 1) + s(-, -, x_{i_3}^3, \ldots, x_{i_n}^n), \\ \vdots \end{cases}$$

where all possible combinations of gaps occur except for the one where all the residues are replaced with gaps. The algorithm is initialized by the condition $F(0, \ldots, 0) = 0$. The traceback begins at the element $F(m_1, \ldots, m_n)$ and is similar to that for pairwise alignments. The matrix $\left(F(i_1, \ldots, i_n) \right)$ with $0 \le i_j \le m_j$, $j = 1, \ldots, n$, is an $(m_1 + 1) \times \cdots \times (m_n + 1)$-matrix, and it is convenient to visualize it by looking at its two-dimensional sections.

Example 3.4 We will find all optimal alignments of the sequences $x = CCGT$, $y = AGT$, $z = CCA$ using the scoring scheme as follows: one calculates the score of an alignment from the scores of its columns M_i's by applying formula (3.2); if M_i has three identical symbols, set $s(M_i) = 2$; if it has exactly two identical symbols, but no gaps, set $s(M_i) = 1$; if it has three distinct symbols, but no gaps, set $s(M_i) = -1$; if it has exactly one gap, set $s(M_i) = -2$; if it has two gaps, set $s(M_i) = -4$. Here the indices i_1, i_2, and i_3 correspond to sequences x, y, and z, respectively.

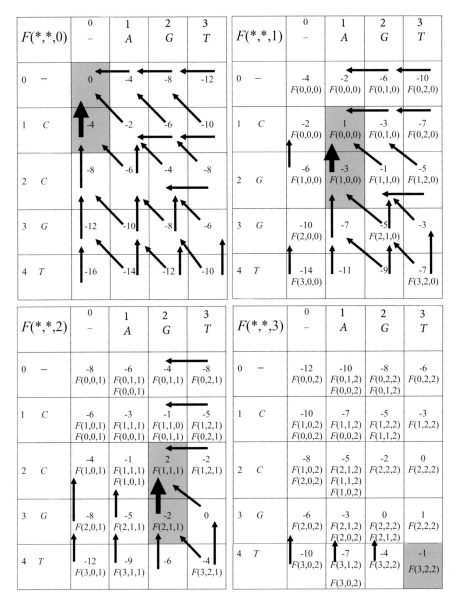

Fig. 3.6 The sections of matrix F in direction i_3

Figure 3.6 gives the sections of F in direction i_3. As before, the arrows and labels show from which elements each number was obtained. The shaded cells and thick arrows indicate the traceback; it begins at $F(4, 3, 3) = -1$ and travels through the shaded cells until it reaches $F(0, 0, 0)$. The traceback leads to the following three paths:

$$F(4, 3, 3) \rightarrow F(3, 2, 2) \rightarrow F(2, 1, 1) \rightarrow F(1, 0, 0) \rightarrow F(0, 0, 0),$$
$$F(4, 3, 3) \rightarrow F(3, 2, 2) \rightarrow F(2, 1, 1) \rightarrow F(1, 1, 1) \rightarrow F(0, 0, 0),$$
$$F(4, 3, 3) \rightarrow F(3, 2, 2) \rightarrow F(2, 2, 2) \rightarrow F(1, 1, 1) \rightarrow F(0, 0, 0).$$

They give rise, respectively, to the following three optimal alignments having score -1:

$$
\begin{array}{lll}
x : C\ C\ G\ T & x : C\ C\ G\ T & x : C\ C\ G\ T \\
y : - A\ G\ T, & y : A - G\ T, & y : A\ G - T \\
z : - C\ C\ A & z : C - C\ A & z : C\ C - A.
\end{array}
$$

Owing to the memory and time complexity, in practice the algorithm discussed above cannot be utilized to align a large number of sequences. As a result, alternative (heuristic) algorithms have been developed. We will briefly mention some of them below.

3.6.1 MSA

MSA builds upon the aforementioned multi-dimensional dynamic programming algorithm, while incorporating certain constraints to mitigate the time complexity.

For simplicity, let us assume an SP-scoring system for both the residues and gaps (i.e., an SP-scoring scheme with linear gap model). Thus, the score of a multiple alignment is the sum of the scores of all the induced pairwise alignments. Denote by M a multiple alignment and by M^{kl} the induced pairwise alignment between sequences k and l. Then

$$\mathscr{S}(M) = \sum_{k < l} \mathscr{S}(M^{kl}),$$

where $\mathscr{S}(M^{kl})$ is the score of M^{kl}. Clearly, if s^{kl} is the score of an optimal global alignment between sequences k and l, then $\mathscr{S}(M^{kl}) \leq s^{kl}$.

Suppose now that there is a lower bound τ on the score of an optimal multiple alignment. This threshold can be established using a rapid heuristic multiple alignment algorithm, such as the forthcoming Star Alignment algorithm we will delve into. Given this context, we can assert the following relationship for the optimal multiple alignment M_0:

$$\tau \leq \mathscr{S}(M_0) = \sum_{k' < l'} \mathscr{S}(M_0^{k'l'}) \leq \mathscr{S}(M_0^{kl}) - s^{kl} + \sum_{k' < l'} s^{k'l'},$$

for all k, l. It then follows that

$$\mathscr{S}(M_0^{kl}) \geq t^{kl},$$

where

$$t^{kl} := \tau + s^{kl} - \sum_{k'<l'} s^{k'l'}.$$

The scores of the pairwise optimal alignments in the right-hand side of this formula can be calculated as discussed in the earlier sections and therefore t^{kl} can be computed.

Thus, we only need to look for the multiple alignments that induce pairwise alignments whose scores are no less than t^{kl}. This observation significantly reduces the number of elements in the multi-dimensional dynamic programming matrix that need to be taken into account and as a result increases the computational speed.

3.6.2 Progressive Alignment

Progressive alignment methods are heuristic techniques that generate multiple alignments from pairwise alignments. Typically, these methods follow this procedure: initially, two sequences are chosen and aligned, then a third sequence is selected and aligned to existing alignment of the first two sequences. This process continues iteratively until all sequences have been incorporated. ClustalW is an illustrative algorithm that falls within this category [33]. It is important to note that multiple alignments obtained through the progressive alignment approach often require manual refinement to ensure accuracy.

We will now describe one simple progressive alignment algorithm, called the Star Alignment algorithm. It is a rather fast heuristic method for producing multiple alignments. Certainly, just like any other heuristic algorithm, it is not guaranteed to find an optimal alignment. The idea of this method is to choose a sequence that has the most similarity to all the other sequences and utilize it as the center of a "star" aligning all the other sequences to it. We will explain the algorithm using the following example.

Example 3.5 Suppose we are given the following DNA sequences:

$$x^1 : CGGATTCGG$$
$$x^2 : CGAATTCGG$$
$$x^3 : CGTTCCGGGG$$
$$x^4 : CGTGGTGG$$
$$x^5 : CTGACTT.$$

We adopt the scoring scheme for pairwise alignments from Example 3.1, utilize the corresponding SP-scoring scheme with linear gap model, compute all the pairwise optimal scores (i.e., the scores found by the global pairwise alignment algorithm outlined in Sect. 3.2), arrange them into the matrix shown below, and find the sum in each row:

	x^1	x^2	x^3	x^4	x^5	Total Score
x^1		7	-2	0	-3	2
x^2	7		-2	0	-4	1
x^3	-2	-2		0	-7	-11
x^4	0	0	0		-3	-3
x^5	-3	-4	-7	-3		$-17.$

Among these sequences, x^1 has the best total score (which is equal to 2) and is chosen to be at the center of the future star. Optimal alignments between x^1 and each of the remaining sequences are found by the global alignment algorithm from Sect. 3.2:

$$x^1 : C\,G\,G\,A\,T\,T\,C\,G\,G$$
$$x^2 : C\,G\,A\,A\,T\,T\,C\,G\,G,$$

$$x^1 : C\,G\,G\,A\,T\,T\,C\,G\,G - -$$
$$x^3 : C\,G\,T - T\,C\,C\,G\,G\,G\,G,$$

$$x^1 : C\,G\,G\,A\,T\,T\,C\,G\,G$$
$$x^4 : C\,G\,T\,G\,G\,T - G\,G,$$

$$x^1 : C\,G\,G\,A\,T\,T\,C\,G\,G$$
$$x^5 : C\,T\,G\,A\,C\,T\,T - -.$$

We will now merge the pairwise alignments displayed above using the "once a gap – always a gap" principle. We begin with x^1 and x^2:

$$x^1 : C\,G\,G\,A\,T\,T\,C\,G\,G$$
$$x^2 : C\,G\,A\,A\,T\,T\,C\,G\,G$$

and add x^3, but, as x^3 is longer than each of x^1 and x^2, we add gaps at the ends of x^1 and x^2:

$$x^1 : C\,G\,G\,A\,T\,T\,C\,G\,G - -$$
$$x^2 : C\,G\,A\,A\,T\,T\,C\,G\,G - -$$
$$x^3 : C\,G\,T - T\,C\,C\,G\,G\,G\,G.$$

These gaps are never deleted. If we introduce a gap at some position, it will stay there until the end of the alignment process.

Now, we add x^4 and x^5. They also must have gaps added to their ends. The resulting multiple alignment is shown below:

$$x^1 : C\ G\ G\ A\ T\ T\ C\ G\ G\ -\ -$$
$$x^2 : C\ G\ A\ A\ T\ T\ C\ G\ G\ -\ -$$
$$x^3 : C\ G\ T\ -\ T\ C\ C\ G\ G\ G\ G$$
$$x^4 : C\ G\ T\ G\ G\ T\ -\ G\ G\ -\ -$$
$$x^5 : C\ T\ G\ A\ C\ T\ T\ -\ -\ -\ -.$$

The alignments found by the Star Alignment algorithm may not be optimal as is the case in the example below.

Example 3.6 Consider the sequences

$$x^1 :\ AATCC$$
$$x^2 :\ ATCTC$$
$$x^3 :\ AACTC$$

and assume the scoring scheme from Example 3.5. Again, here x^1 is the center of the star, and some of the optimal alignments between x^1 and each of the other two sequences are

$$x^1 : A\ A\ T\ C\ -\ C$$
$$x^2 : A\ -\ T\ C\ T\ C,$$

$$x^1 : A\ A\ -\ T\ C\ C$$
$$x^3 : A\ A\ C\ T\ C\ -.$$

Merging these alignments, we obtain

$$x^1 : A\ A\ -\ T\ C\ -\ C$$
$$x^2 : A\ -\ -\ T\ C\ T\ C$$
$$x^3 : A\ A\ C\ T\ C\ -\ -.$$

The above alignment is not optimal. Indeed, its score is equal to -5, whereas a higher-scoring alignment (with score 1) is

$$x^1 : A\ A\ T\ C\ -\ C$$
$$x^2 : A\ T\ -\ C\ T\ C$$
$$x^3 : A\ A\ -\ C\ T\ C.$$

Chapter 4
The Time-Frequency Spectral Analysis and Applications in Bioinformatics

4.1 Introduction

In signal processing, time-frequency analysis encompasses a set of techniques that study a signal across both time and frequency domains concurrently, employing diverse time-frequency representations. The notion that a time series exhibits repetitive or predictable behaviors over time holds pivotal significance, distinguishing time series analysis from classical statistics analyses. The consistency within a time series finds expression in the frequency domain through the identification of periods inherent to the underlying phenomenon governing the series. This periodicity is encapsulated by Fourier frequencies, derived from combinations of sine and cosine functions. Hence, we can regard a time series as an outcome of a system responding to varied driving frequencies. In terms of regression, the time-domain methodology may be likened to a regression of the present state against past occurrences. Conversely, the frequency-domain approach involves a regression of the present state against periodic sine and cosine components [34]. Since biological signals have changing frequency characteristics, time-frequency analysis has a broad scope of applications in bioinformatics. We will explore the application of the Fourier transform in exon prediction study.

4.2 Discrete Fourier Transform

Fourier analysis has become the most valuable tool in spectral data analysis and has consequently been applied to different kinds of data in many scientific or engineering disciplines. This process can be conceptualized as the expansion of a signal $x(n)$ into a collection of waves, typically represented as sine and cosine functions. The Discrete Fourier Transform (DFT) serves as an adaptation of the traditional Fourier analysis. It operates on discrete signals within the time

S. S.-T. Yau et al., *Mathematical Principles in Bioinformatics*, Interdisciplinary Applied Mathematics 58, https://doi.org/10.1007/978-3-031-48295-3_4

domain, transforming them into a series of values in the frequency domain. This transformation is particularly applicable since dealing with discrete signals is more common.

Prior to introducing the Discrete Fourier Transform (DFT), let us first acquaint ourselves with the Euler identity, presented as follows:

$$e^{ix} = \cos x + i \sin x, \tag{4.1}$$

where $i = \sqrt{-1}$. (We assume that readers have some basic knowledge of complex numbers.) This formula enables us to utilize exponential functions to replace sine and cosine functions, yielding:

$$\cos x = \frac{e^{ix} + e^{-ix}}{2}$$
$$\sin x = \frac{e^{ix} - e^{-ix}}{2i}. \tag{4.2}$$

There are numerous advantages to employing this technique. For instance, the computation of exponential functions becomes considerably simpler and the formulation is more unified in this way. Let us now proceed to define discrete Fourier transform (DFT) and inverse discrete Fourier transform (IDFT) using the framework of exponential functions.

Definition 4.1 Given a time-domain sequence $x = (x(0), x(1), \ldots, x(N-1))$ in C^N, we define the discrete Fourier transform (DFT) to be

$$X(k) = \sum_{n=0}^{N-1} x(n) e^{-\frac{2\pi}{N} ikn} \tag{4.3}$$

for $k = 0, 1, \ldots, N-1$.

Definition 4.2 Given a frequency-domain sequence $X = (X(0), X(1), \ldots, X(N-1))$ in C^N, the inverse discrete Fourier transform (IDFT) is defined as

$$x(n) = \frac{1}{N} \sum_{k=0}^{N-1} X(k) e^{\frac{2\pi}{N} ikn} \tag{4.4}$$

for $n = 0, 1, \ldots, N-1$.

The two definitions appear quite similar, differing only in the sign within the exponential functiona and the coefficient $\frac{1}{N}$. We will subsequently show the reason why DFT and IDFT are considered "inverse" operations. To begin with, we will establish the following orthogonality property:

$$\sum_{n=0}^{N-1} e^{\frac{2\pi}{N}i(m-k)n} = \begin{cases} N, & \text{for} \quad m = k \\ 0, & \text{for} \quad m \neq k, \end{cases} \tag{4.5}$$

where $m, k = 0, 1, \ldots, N - 1$. If $m = k$, the summation is equal to N as $e^{\frac{2\pi}{N}i(m-k)n} = e^0 = 1$. Otherwise, by using the summation formula of geometric progression, we get

$$\sum_{n=0}^{N-1} e^{\frac{2\pi}{N}i(m-k)n} = \frac{1 - e^{2\pi i(m-k)}}{1 - e^{\frac{2\pi}{N}i(m-k)}} = 0, \quad \text{for} \quad m \neq k. \tag{4.6}$$

In order to construct $x(n)$ in terms of $X(k)$, we multiply $X(k)$ by $e^{\frac{2\pi}{N}ikn}$ and sum them over the interval $k = 0$ to $k = N - 1$. Note that,

$$\sum_{k=0}^{N-1} X(k)e^{\frac{2\pi}{N}ikn} = \sum_{k=0}^{N-1}\left(\sum_{m=0}^{N-1} x(m)e^{-\frac{2\pi}{N}imk}\right)e^{\frac{2\pi}{N}ikn}$$

$$= \sum_{m=0}^{N-1} x(m)\sum_{k=0}^{N-1} e^{-\frac{2\pi}{N}(m-n)k} = Nx(n). \tag{4.7}$$

Therefore, we show that $x(n)$ is the IDFT of $X(k)$ if $X(k)$ is the DFT of $x(n)$, and vice versa. We use $x(n) \Leftrightarrow X(k)$ to represent this relationship.

Let us compute an example. Assume that $x = (2, 3, 1, 4)$, the DFT of x is computed as

$$X(0) = 2e^{-\frac{0\pi}{4}i} + 3e^{-\frac{0\pi}{4}i} + e^{-\frac{0\pi}{4}i} + 4e^{-\frac{0\pi}{4}i} = 2 + 3 + 1 + 4 = 10$$

$$X(1) = 2e^{-\frac{0\pi}{4}i} + 3e^{-\frac{2\pi}{4}i} + e^{-\frac{4\pi}{4}i} + 4e^{-\frac{6\pi}{4}i} = 2 + 3(-i) + (-1) + 4i = 1 + i$$

$$X(2) = 2e^{-\frac{0\pi}{4}i} + 3e^{-\frac{4\pi}{4}i} + e^{-\frac{8\pi}{4}i} + 4e^{-\frac{12\pi}{4}i} = 2 + 3(-1) + 1 + 4(-1) = -4$$

$$X(3) = 2e^{-\frac{0\pi}{4}i} + 3e^{-\frac{6\pi}{4}i} + e^{-\frac{12\pi}{4}i} + 4e^{-\frac{18\pi}{4}i} = 2 + 3i + (-1) + 4(-i) = 1 - i.$$

Then the DFT result of the sample sequence x is $X = (10, 1 + i, -4, 1 - i)$. On the other hand, the IDFT of X gives the original data x which could be computed as

$$x(0) = \frac{1}{4}\left[10e^{\frac{0\pi}{4}i} + (1 + i)e^{\frac{0\pi}{4}i} + (-4)e^{\frac{0\pi}{4}i} + (1 - i)e^{\frac{0\pi}{4}i}\right]$$

$$= \frac{1}{4}[10 + (1 + i) - 4 + (1 - i)] = 2$$

$$x(1) = \frac{1}{4}\left[10e^{\frac{0\pi}{4}i} + (1+i)e^{\frac{2\pi}{4}i} + (-4)e^{\frac{4\pi}{4}i} + (1-i)e^{\frac{6\pi}{4}i}\right]$$

$$= \frac{1}{4}[10 + (1+i)i + 4 + (1-i)(-i)] = 3$$

$$x(2) = \frac{1}{4}\left[10e^{\frac{0\pi}{4}i} + (1+i)e^{\frac{4\pi}{4}i} + (-4)e^{\frac{8\pi}{4}i} + (1-i)e^{\frac{12\pi}{4}i}\right]$$

$$= \frac{1}{4}[10 + (1+i)(-1) - 4 + (1-i)(-1)] = 1$$

$$x(3) = \frac{1}{4}\left[10e^{\frac{0\pi}{4}i} + (1+i)e^{\frac{6\pi}{4}i} + (-4)e^{\frac{12\pi}{4}i} + (1-i)e^{\frac{18\pi}{4}i}\right]$$

$$= \frac{1}{4}[10 + (1+i)(-i) + 4 + (1-i)i] = 4.$$

Since the sequence in the frequency domain can be complex even if the sequence in the time domain is real, we sometimes use DFT power spectrum PS to represent the sequence in the frequency domain where

$$PS(k) = |X(k)|^2.$$

Two important properties of DFT are applied in the following content. The first property is the conjugate relation between $X(k)$ and $X(N-k)$ when the sequence in the time domain is real. Given a sequence $x = (x(0), \ldots, x(N-1))$ in $\mathbb{R}^N$,

$$X(N-k) = \sum_{n=0}^{N-1} x(n)e^{-\frac{2\pi}{N}i(N-k)n}$$

$$= \sum_{n=0}^{N-1} x(n)e^{\frac{2\pi}{N}ikn}$$

$$= \left(\sum_{n=0}^{N-1} x(n)e^{-\frac{2\pi}{N}ikn}\right)^*$$

$$= X(k)^*, \tag{4.8}$$

where $*$ is the conjugate operator, that is, $(a+bi)^* = a - bi$ for real numbers a, b. Therefore, we can obtain $X(N-k)$ by $X(k)$. As a direct result, we have

$$PS(k) = PS(N-k). \tag{4.9}$$

It is the reason why we only present half of the picture in the frequency domain. Another significant property is the well-known Parseval's theorem.

Theorem 4.1 *Let $x(n) \Leftrightarrow X(k)$ with sequence length N, then*

$$\sum_{n=0}^{N-1} |x(n)|^2 = \frac{1}{N} \sum_{k=0}^{N-1} |X(k)|^2. \tag{4.10}$$

Proof The proof of the theorem is direct:

$$\sum_{n=0}^{N-1} |x(n)|^2 = \sum_{n=0}^{N-1} x(n)x^*(n)$$

$$= \frac{1}{N^2} \sum_{n=0}^{N-1} \left(\sum_{k=0}^{N-1} X(k)e^{\frac{2\pi}{N}ikn}\right)\left(\sum_{m=0}^{N-1} X(m)e^{\frac{2\pi}{N}imn}\right)^*$$

$$= \frac{1}{N^2} \sum_{n=0}^{N-1}\sum_{k=0}^{N-1}\sum_{m=0}^{N-1} X(k)X^*(m)e^{\frac{2\pi}{N}i(k-m)n} \tag{4.11}$$

$$= \frac{1}{N^2} \sum_{k=0}^{N-1}\sum_{m=0}^{N-1} X(k)X^*(m) \sum_{n=0}^{N-1} e^{\frac{2\pi}{N}i(k-m)n}$$

$$= \frac{1}{N} \sum_{k=0}^{N-1} |X(k)|^2.$$

The quantity $\sum_{n=0}^{N-1} |x(n)|^2$ can be referred to as the energy in the time domain, while $\frac{1}{N} \sum_{k=0}^{N-1} |X(k)|^2$ can be interpreted as the energy in the frequency domain. This theorem demonstrates the equality between these two energy functions. Furthermore, it implies that the Euclidean distance between two signals $x(n)$ and $y(n)$ in the time domain is the same as their Euclidean distance in the frequency domain divided by N. Sometimes we may use DFT power spectrum to index and query signals in databases and compare the difference between two signals in the frequency domain.

Fourier transforms are frequently employed to identify the frequency components of a signal that are obscured within a noisy time-domain signal. As an example, consider the following signal containing a 50 Hz sinusoid of amplitude 1.5, 120 Hz sinusoid of amplitude 1, and 240 Hz cosine of amplitude 3 and is corrupted with some zero-mean random noise.

$$s = 1.5sin(2 \cdot \pi \cdot 50 \cdot t) + sin(2 \cdot \pi \cdot 120 \cdot t) + 3cos(2 \cdot \pi \cdot 240 \cdot t) + random \tag{4.12}$$

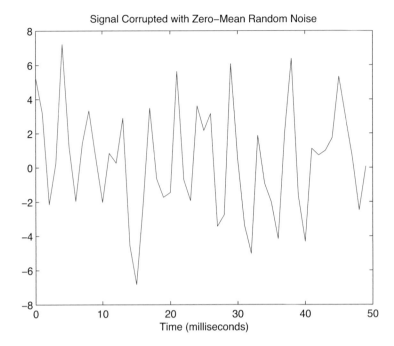

Fig. 4.1 Plotting the test signal vs time t

It is difficult to identify the three frequency components, 50, 120, and 240 Hz in the original time-domain signal (Fig. 4.1), but after converting to the frequency domain by the discrete Fourier transform, these three frequency components can be detected after the signal is converted to the frequency domain by the DFT (Fig. 4.2).

From this example, we see that the signal is made up of three different colors (frequencies) at various strengths (amplitudes), then we might consider the power spectrum periodogram as a prism that decomposes the color into its primary colors. Hence we call the frequency analysis spectral analysis.

4.3   Exon Prediction Based on Fourier Spectral Analysis

4.3.1   Eukaryotic Gene Structure

Let us briefly revisit some of the concepts we covered in Chap. 1. Living cells can be classified into two categories: prokaryotes, such as bacteria, in which the cells do not have a distinct nucleus, and eukaryotes, such as most animal cells, in which the cells have distinct nuclei. For eukaryote cells, genes are composed of alternating stretches of exons (coding regions) and introns (non-coding regions).

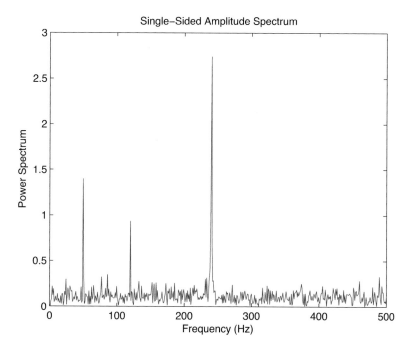

Fig. 4.2 Plotting DFT power spectrum vs frequency

During transcription, both exons and introns are transcribed into pre-mRNA in their linear order. Thereafter, a process called splicing takes place, in which the intron sequences are excised and discarded from the RNA sequence. The remaining RNA segments, the ones corresponding to the exons, are ligated to form the mature RNA strand.

Distinguishing exons from DNA sequences is a significant challenge. This difficulty arises due to the absence of distinctive characteristics between exons and introns in the primary DNA sequence. To address this issue, numerous computational methods such as hidden Markov models have been proposed [35]. Nevertheless, there remains a desire to uncover intrinsic features of exons for the development of more natural techniques. In this section, we will explore the application of DFT to solve this problem.

4.3.2 Fourier Spectrum Analysis of DNA Sequences

A DNA sequence can be depicted as a permutation of four characters (A, T, C, and G) of varying lengths. However, this representation is not directly suitable for DFT. To accommodate this, binary indicator sequences are introduced. Let us take a DNA sequence denoted as $x = (x(0), x(1), \ldots, x(N-1))$ as an example. This sequence can be decomposed into four binary indicator sequences: $u_A(n)$, $u_T(n)$, $u_C(n)$, and $u_G(n)$. These indicators represent the presence or absence of each nucleotide (A, T, C, and G) at the nth position, respectively. For example, the indicator sequence $u_A(n) = 0001010111 \ldots$ indicates that the nucleotide A presents in positions 4, 6, 8, 9, and 10 of the DNA sequence. (The position four corresponds to $u_A(3)$ since the index starts from 0.) That is,

$$U_\alpha(n) = \begin{cases} 1, & x(n) = \alpha \\ 0, & \text{otherwise,} \end{cases} \qquad (4.13)$$

where $\alpha \in \{A, T, C, G\}$, $n = 0, 1, \ldots, N - 1$.

According to the Definition 4.1, we define the DFT of the indicator sequences of the DNA sequence x to be

$$U_\alpha(k) = \sum_{n=0}^{N-1} u_\alpha(n) e^{-\frac{2\pi}{N} ikn} \qquad (4.14)$$

for $\alpha \in \{A, T, C, G\}$, $k = 0, 1, \ldots, N - 1$ and we define the DFT power spectrum of a binary indicator at the frequency k to be:

$$PS_\alpha(k) = |U_\alpha(k)|^2, \quad k = 0, 1, \ldots, N - 1. \qquad (4.15)$$

The DFT power spectrum of a DNA sequence is the sum of the power spectrum of its four binary indicator sequences:

$$PS(k) = PS_A(k) + PS_T(k) + PS_C(k) + PS_G(k), \qquad (4.16)$$

where $PS_A(k)$, $PS_T(k)$, $PS_C(k)$, and $PS_G(k)$ are the Fourier power spectrum of the four indicator sequences $u_A(n)$, $u_T(n)$, $u_C(n)$, and $u_G(n)$, respectively.

Based on Parseval's Theorem, the energy in the time domain $\sum_{n=0}^{N-1} |u_\alpha(n)|^2$ equals to the energy in the frequency domain $\frac{1}{N} \sum_{k=0}^{N-1} |U_\alpha(k)|^2$. (In Fourier analysis, the term "time domain" is a specialized term. However, when dealing with sequences, it is not necessarily associated with time. In this context, it can be better understood as the spatial domain.) Therefore, by Parseval's theorem and the fact that

$$|u_A(n)|^2 + |u_T(n)|^2 + |u_C(n)|^2 + |u_G(n)|^2 = 1, \; \forall n, \qquad (4.17)$$

we have

$$E = \frac{1}{N} \sum_{k=0}^{N-1} PS(k)$$

$$= \frac{1}{N} \sum_{k=0}^{N-1} (|U_A(k)|^2 + |U_T(k)|^2 + |U_C(k)|^2 + |U_G(k)|^2) \tag{4.18}$$

$$= \sum_{n=0}^{N-1} (|u_A(n)|^2 + |u_T(n)|^2 + |u_C(n)|^2 + |u_G(n)|^2)$$

$$= N,$$

where E is the total energy of four indicator sequences in frequency domain, which can be considered as noise background in gene-finding methods.

4.3.3 The 3-Base Periodicity in Exon Sequences

Over the past two decades, a range of computational algorithms have emerged for exon prediction. The predominant approach in exon finding methods is rooted in statistical techniques, often utilizing training datasets of known exon and intron sequences to derive prediction functions. For instance, the GenScan algorithm assesses distinct statistical features of exons and introns within a genome, employing them in prediction through a hidden Markov model (HMM) [35]. Another method, MZEF, is grounded in quadratic discriminant analysis of diverse sequence characteristics in exons and introns [36]. However, statistical methods rely on training sets from which various statistical parameters are derived, rendering them less effective in scenarios where training data are unavailable. Consequently, the pursuit of gene prediction techniques that transcend statistical approaches has emerged as a fundamental endeavor in gene-finding research [37].

In recent years, signal processing methodologies, particularly time-frequency analysis, have garnered considerable attention in genomic DNA research. They offer the potential to unveil concealed periodicities within sequences, thus proving increasingly valuable in unraveling genome structures. By transforming symbolic DNA sequences into numerical counterparts, signal processing tools like the Fourier transform or wavelet analysis can be applied to the numerical vectors, enabling the exploration of sequence frequency domains [38, 39]. These investigations are based on a crucial characteristic of exon sequences: the 3-base periodicity, which is recognized by a distinct peak at the frequency $N/3$ within the Fourier power spectrum of the DNA sequence, where N signifies the sequence length.

We take a simple example to show why the peak at the frequency $N/3$ is related to the 3-base periodicity of the sequence in the time domain for DFT. Consider a sequence with period 3 and length $N = 3M$:

$$x(t) = \begin{cases} 1, & t = 0 \ (mod \ 3) \\ 0, & otherwise \end{cases} \qquad (4.19)$$

Then

$$X(k) = \sum_{n=0}^{3M-1} x(n) e^{-\frac{2\pi}{3M} ikn}$$

$$= \sum_{m=0}^{M-1} e^{-\frac{2\pi}{M} ikm} \qquad (4.20)$$

$$= \begin{cases} 0 & , \ \frac{k}{M} \notin Z \\ M & , \ \frac{k}{M} \in Z \end{cases}$$

and

$$PS(k) = \begin{cases} 0 & , \ \frac{k}{M} \notin Z \\ M^2 & , \ \frac{k}{M} \in Z \end{cases} \qquad (4.21)$$

Hence, it becomes evident that the 3-base periodicity of the sequence results in a peak at the frequency $M = N/3$. However, in real-world scenarios, sequences exhibiting the 3-base periodicity might not be as regular as the illustrative example above. Consequently, there is a large possibility that $X(k)$ may not be zero outside the peak frequency.

Research has shown that the 3-base periodicity is prevalent in most exon sequences, but not in intron sequences [38]. The existence of the 3-base periodicity is the basis of the Fourier exon prediction method. Figure 4.3 is the DFT power spectrum of an exon (a) and an intron (b) from the gene of a fruit fly. (As mentioned before, we only show the first half of the DFT spectrum.) It is easy to see a peak at the frequency $\frac{N}{3}$ for exons but not for introns.

4.3.4   PS(N/3) Is Determined by the Unbalanced Nucleotide Distributions of the Three Codon Positions

The computation time for Fourier transforms is expensive, especially for long DNA sequences [40, 41]. However, for the gene prediction problem, we only need to calculate $PS(\frac{N}{3})$, which is determined by the unbalanced nucleotide distributions of the three codon positions [40, 43–46].

Given a nucleotide sequence $(x(0), x(1), \ldots, x(N-1))$ $(N = 3M)$, we can divide the nucleotides into three classes: the first codon positions $(x(0), x(3), \ldots, x(N-3))$, the second codon positions $(x(1), x(4), \ldots, x(N-2))$,

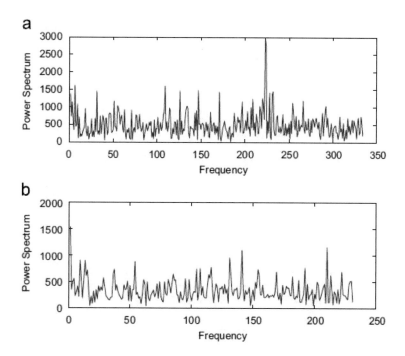

Fig. 4.3 DFT power spectrum of an exon (**a**) and an intron (**b**) from the gene of AAM70818.2 of *Drosophila melanogaster (fruit fly)*

and the third codon positions $(x(2), x(5), \ldots, x(N - 1))$. In exonic regions, an imbalance has been noted in the nucleotide distributions across the three classes. In contrast, these distributions are balanced within intronic regions. The occurrence of this unbalanced distribution in the three classes can be attributed to redundant mappings within coding amino acids, coupled with the preference of proteins for specific amino acid compositions.

To investigate the relationship between the nucleotide distributions and the 3-base periodicity of a DNA sequence, we calculate the number of occurrences of different nucleotides of three classes and denote them as F_{Ai}, F_{Ti}, F_{Ci}, and F_{Gi} where $i = 1, 2, 3$ for the first, second, third codon positions, respectively. In other words, we can get the following matrix:

$$F = \begin{pmatrix} F_{A1} & F_{A2} & F_{A3} \\ F_{T1} & F_{T2} & F_{T3} \\ F_{C1} & F_{C2} & F_{C3} \\ F_{G1} & F_{G2} & F_{G3} \end{pmatrix}.$$

In fact, $PS(\frac{N}{3})$ is determined by this matrix.
For $x \in \{A, T, C, G\}$,

$$U_x\left(\frac{N}{3}\right) = \sum_{n=0}^{N-1} u_x(n) e^{-\frac{2\pi}{3} in}$$

(4.22)

$$= F_{x1} + e^{-\frac{2\pi}{3}i} F_{x2} + e^{-\frac{4\pi}{3}i} F_{x3}.$$

$$PS_x\left(\frac{N}{3}\right) = (F_{x1} - \frac{1}{2} F_{x2} - \frac{1}{2} F_{x3})^2 + \frac{3}{4}(F_{x2} - F_{x3})^2$$

$$= F_{x1}^2 + F_{x2}^2 + F_{x3}^2 - F_{x1}F_{x2} - F_{x1}F_{x3} - F_{x2}F_{x3}.$$

(4.23)

Therefore,

$$PS\left(\frac{N}{3}\right) = \sum_{x=A,T,C,G} (F_{x1}^2 + F_{x2}^2 + F_{x3}^2 - F_{x1}F_{x2} - F_{x1}F_{x3} - F_{x2}F_{x3}).$$

(4.24)

Since the matrix F is relatively easy to calculate, the computation for $PS(\frac{N}{3})$ can be simplified by the formula (4.24).

We can reformulate the formula (4.24) and get

$$PS\left(\frac{N}{3}\right) = \frac{3}{2} \sum_{x=A,T,C,G} \sum_{i=1,2,3} \left(F_{xi} - \frac{1}{3} \sum_{j=1,2,3} F_{xj}\right)^2,$$

(4.25)

where the right-hand side can be regarded as a measurement of the unbalance of the distributions among three classes.

4.3.5 Algorithm for Finding Exons by Nucleotide Distribution (FEND)

The background noise of a DNA sequence of length N, represented by E, the total energy of four indicator sequences in frequency domain, is proved to be N. Thus, the ratio of the 3-base periodicity signal to the background noise of a DNA sequence, denoted as $SN(N)$, is defined as follows:

$$SN(N) = \frac{PS(\frac{N}{3})}{N}.$$

(4.26)

$SN(N)$ represents the measure of the strength of the 3-base periodicity per nucleotide in a given DNA sequence. This characteristic has been observed to differentiate between exon and intron sequences. Typically, most exon sequences exhibit $SN(N)$ values equal to or greater than 2, whereas most intron sequences tend to have $SN(N)$ values less than 2. Therefore, by evaluating the value of $SN(N)$, it is possible to determine whether a sequence corresponds to an exon or an intron.

However, in real applications, the problem is much more complicated. Real DNA sequences are composed of both exons and introns, rather than consisting solely of either type. Therefore, the challenge lies in accurately identifying the exonic nucleotides within long sequences. This task cannot be effectively accomplished solely by computing $SN(N)$ and necessitates the development of a new algorithm to achieve accurate results.

One way to solve this problem is to consider sliding windows of a fixed length [38], which break the long sequence into subsequences of the window length. For this method, the window length should be selected properly. A small window length emphasizes the small peaks that appear due to the background noise and a large window length causes short exons or introns in DNA sequences to be missed.

In this part, we will focus on another method to deal with the problem, which is to consider the slope of SN. We first introduce some notations used in the following part. For a DNA sequence of length N, let D_k denotes the sub-region ranging from the beginning to the position k (also called DNA walk of length k). In addition, since the sub-region changes in the algorithm, to avoid the ambiguity for $PS(\frac{N}{3})$ (not only related to the position but also related to the sub-region), we use $PS(D_k, \frac{k}{3})$ and $SN(D_k)$ to denote the 3-base periodicity signal and the ratio of 3-base periodicity signal to the background noise of D_k, respectively.

It is found that $SN(D_k)$ has different trends when k increases between exons and introns. In Fig. 4.4a we plot the average $SN(k)$ of DNA walks of 1000 base pairs fragment of 258 exon sequences from the human genome. In Fig. 4.4b we plot the average $SN(k)$ of DNA walks of 1000 base pairs fragment of 216 intron sequences from the human genome. The result is that $SN(D_k)$ increases as k increases for a pure exon sequence while it randomly fluctuates around some low values as k increases for a pure intron sequence.

This phenomenon inspires us to consider the slope of SN. The algorithm for finding exons by nucleotide distribution (or FEND) is developed as follows [42]. (Fig. 4.5 depicts the flowchart of the algorithm.)

1. Let $k = 1$.
2. Calculate the nucleotide distributions within the three codon positions of D_k (represented by the F matrix denoted as F_k). The matrix F_k is recursively derived from the previous matrix F_{k-1} and the nucleotides at position k when $k > 1$.
3. Compute $SN(D_k)$ by formula (4.24) and formula (4.26).
4. Increment k by 1 and iterate through steps 2 to 3 until $k = N$.
5. Calculate the slope of SN at each position on the SN plot by the following way: Given that the majority of exon or intron sequences within a genome surpass 50 base pairs in length, the slope at the ith position is determined as $(SN(D_i) - SN(D_{i-50}))/50$, with i ranging from 51 to N.
6. Assign the nucleotide at each position to either the exon or intron region based on the following criteria: If the slope at the position is greater than 0 and SN is equal to or greater than 2, designate the nucleotide at that position as an exon nucleotide. Otherwise, classify it as an intron nucleotide.

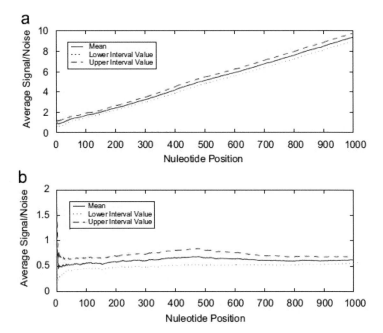

Fig. 4.4 Plots depicting average signal-to-noise ratios of DNA walks for 1000 bp DNA fragments from the human genome are shown. (The average signal-to-noise ratios are in the middle and the corresponding 95% confidence intervals of these average ratios are illustrated in the lower and upper plots.) These fragments comprise 258 exons and 216 introns. (**a**) Exons. (**b**) Introns

7. If a DNA region with fewer than 50 base pairs is categorized as an intron in step 6, and it is enclosed by two exon regions, this region is commonly a false negative and should be reclassified as an exon region. Correspondingly, if a DNA region with less than 50 base pairs is identified as an exon in step 6 and is surrounded by two intron regions, it is often a false positive and should be reassigned as an intron region.

In the case of a lengthy DNA sequence that might encompass more than two exons (or introns), such as an exon-intron-exon arrangement, the cumulative signal-to-noise ratio of the final exon could decline, particularly when a lengthy intron is positioned between them. This potential decline could influence the prediction accuracy. Enhancements to the algorithm could be made by segmenting a DNA sequence into distinct sub-regions. Moreover, in order to diminish the occurrence of false exons and false introns, the algorithm is applied from various arbitrary starting points, allowing multiple evaluations for each nucleotide. The subsequent algorithm has been devised to enhance the precision of exon prediction when utilizing the FEND method:

1. If a DNA sequence is longer than 2000 base pairs (bp), divide it into subsequences of 2000 base pairs.

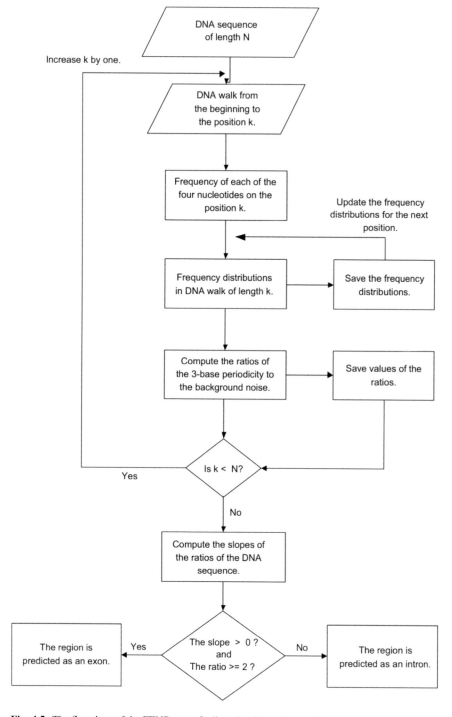

Fig. 4.5 The flowchart of the FEND exon finding algorithm (Step 1–6)

2. For each 2000 base pairs sub-sequence, set $P_1 = 1$, $P_2 = 401$, $P_3 = 801$, $P_4 = 1201$, $P_5 = 1601$, and $P_6 = 2000$ be the six even-spaced points.
3. Identify exon or intron nucleotides using the FEND method on the sub-sequence between point P_i and P_6 where $i = 1, 2, 3, 4, 5$. So each nucleotide after points P_3 is tested at least three times using the FEND method from different start points. A nucleotide is identified as an exon nucleotide when it is predicated in an exon region in the majority of the tests.

The performance evaluation of the FEND algorithm involves the metrics of sensitivity, specificity, and accuracy, which are defined as follows: Sensitivity (S_n): Sensitivity is calculated using the formula $S_n = \frac{TP}{TP+FN}$. Here, TP represents the true positive, which corresponds to the nucleotide length of correctly predicted exons; FN corresponds to the false negative, representing the nucleotide length of incorrectly predicted introns. Specificity (S_p): Specificity is computed using the formula $S_p = \frac{TN}{TN+FP}$. In this equation, TN stands for the true negative, which is the nucleotide length of correctly predicted introns; FP represents the false positive, indicating the nucleotide length of incorrectly predicted exons. Accuracy (AC): The accuracy is defined as the average of sensitivity and specificity, expressed as $AC = \frac{S_n+S_p}{2}$. In summary, sensitivity (S_n) captures the correct prediction of coding sequences, specificity (S_p) measures the accurate prediction of non-coding sequences, and accuracy (AC) represents the overall performance, taking into account both sensitivity and specificity.

To assess the FEND algorithm's viability for predicting protein-coding regions, we employ test datasets comprising full-length gene sequences that encompass both introns and exons. The FEND algorithm is then applied to these full-length gene sequences to validate its efficacy. As an illustrative instance, Fig. 4.6a presents the $SN(D_k)$ plot for a test gene exhibiting an exon-intron-exon structure (gene locus: AAB26989.1 of Drosophila melanogaster). Figure 4.6b depicts the slope plot derived from SN, revealing predominantly positive slopes in exon regions and negative slopes in introns. Figure 4.6c showcases expected gene structures verified by biological experiments. Figure 4.6d displays predicted gene structures without applying the enhanced algorithm. For this case, the values of S_n, S_p, and AC are 0.8684, 0.4372, and 0.6528, respectively. Upon implementing the improved algorithm, Fig. 4.6e portrays the predicted gene structure. The ensuing values of S_n, S_p, and AC are 0.9450, 0.7556, and 0.8503, respectively. Notably, the FEND algorithm's accuracy (AC) enhances by 19.75% in this particular test case. These outcomes underscore the algorithm's capability in effectively identifying the majority of exon and intron sequences, particularly following the algorithmic enhancement.

The FEND algorithm has the following advantages: (1) It uses extendable windows to compute the 3-base periodicity, which reduces the bias when fixed window lengths are used. (2) The computation of magnitude of the 3-base periodicity is based on nucleotide distributions on the three coding positions. The computation of the nucleotide distributions on the DNA walk sequences uses a recursive approach in which the computation of nucleotide distributions on the DNA sequence of length

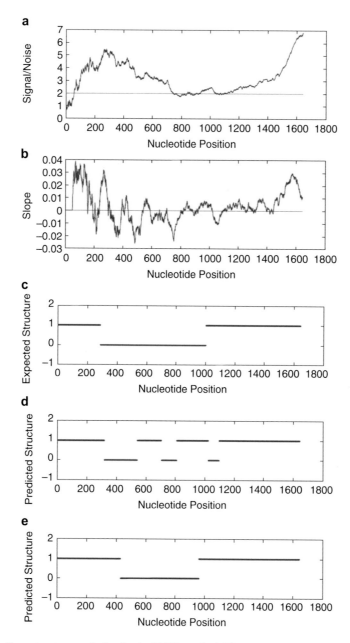

Fig. 4.6 Gene structure prediction by the FEND method. The gene locus is at AAB26989.1 of Drosophila melanogaster (fruit fly). (**a**) The signal-to-noise ratios SN of the DNA walk from this sequence calculated by the FEND method. (**b**) The plot of the slopes of every two points at a distance of 50 base pairs from the SN plot. (**c**) The expected gene structure that is verified by biological experiments. Exon regions are marked as 1, and intron regions are marked as 0. (**d**) The predicted gene structure by the FEND method without improvement. (**e**) The predicted gene structure by the improved FEND method

k uses the results of the nucleotide distributions on the $k - 1$ length DNA segment. In terms of computational complexity, the algorithm has a linear computation time proportional to the length of the DNA sequence, which is very efficient. (3) This method does not require training data sets as in statistical methods. Thus it is an ab initio method, which is very useful when information on the known gene structure is limited.

4.4   DNA Comparison Based on Fourier Spectral Analysis

Comparing DNA sequences, which is the basis of clustering and classification, is of great significance in Bioinformatics. Sequence alignment introduced in Chap. 3 is a good method to compare the sequences but it is too time-consuming for some large-scale problems. Therefore, the alignment-free sequence comparisons attract much attention.

A basic approach for alignment-free sequence comparisons is developing a method to transform sequences into vectors. Then we can compare two sequences by calculating the distance between their corresponding vectors. In this section, we propose several methods that can form vectors from DNA sequences based on DFT.

4.4.1   Even Scaling Method of Fourier Power Spectrum

The power spectrum obtained from the Discrete Fourier Transform (DFT) of a DNA sequence results in a vector, the length of which is contingent upon the sequence's length. Consequently, comparing the power spectra of two sequences with distinct lengths directly is not feasible. A previous solution involved truncating the vectors and utilizing partial spectra, but this approach risked losing information crucial for sequence comparison. To surmount this challenge, we propose an even scaling method, outlined in our previous work [49], aimed at aligning DFT power spectra of varying lengths to a uniform length. This scaling technique is versatile and applicable to diverse data series. In this context, we will refrain from explicating the method through the lens of the DFT process.

Let $T_n(1), \ldots, T_n(n)$ denote a data series. Our goal is to stretch it into a data series of length m, denoted by $T_m(1), \ldots, T_m(m)$. Let $Q(k) = \frac{kn}{m}$ and $R(k) = \max\{1, \lfloor \frac{kn}{m} \rfloor\}$ where the symbol $\lfloor \ldots \rfloor$ denotes the floor function. The even scaling operation on the original power spectrum T_n to T_m is defined as follows:

$$T_m(k) = \begin{cases} T_n(Q(k)), & Q(k) \in \mathbb{Z}^+, \\ T_n(R(k)) + (Q(k) - R(k))(T_n(R(k)+1) - T_n(R(k))), & Q(k) \notin \mathbb{Z}^+ \end{cases}$$

$$\tag{4.27}$$

By this method, we can stretch the DFT spectrum to a given length m. In real applications, m is determined according to the longest length of the DNA sequences in a data set. It is worth noticing that we always exclude the zeroth term in the power spectrum because it is just the sum of data and its value is too large compared with other terms.

Having standardized the DNA sequence vectors to uniform lengths, we can proceed to compute the distance between two vectors for the purpose of comparing their corresponding sequences by the Euclidean distance shown as follows:

$$d((x_1, \ldots, x_n), (y_1, \ldots, y_n)) = \sqrt{\sum_{k=1}^{n}(x_k - y_k)^2}. \tag{4.28}$$

By this distance, we can construct the phylogenetic tree of the genomes or make a classification of the genomes.

4.4.2   Power Spectrum Moment Method

In this part, we consider the moments of the power spectra to deal with the problem of different lengths. In other words, for nucleotide A we can define its j-th moment to be

$$M_j^A = a_j^A \sum_{k=0}^{N-1}(PS_A(k))^j, \quad j = 1, 2, \ldots, \tag{4.29}$$

where a_j^A is a scaling factors [51]. Our objective is to achieve the convergence of higher moments toward zero, ensuring that essential information is retained primarily within the initial moments. As a result, the selection of normalization factors a_j^A should align with the inherent characteristics of the sequences.

By Parseval theorem, we have

$$\sum_{k=0}^{N-1} PS_A(k) = N_A N, \tag{4.30}$$

where N_A is the number of 1 in the binary sequence u_A. So it is reasonable for a_j^A to be a power of $N_A N$. As stated above, we want moments to converge to zero gradually so that information loss is minimal, thus $a_j^A = 1/(N_A N)^{j-1}$ will be a good choice, i.e., we have

$$M_j^A = \frac{1}{N_A^{j-1} N^{j-1}} \sum_{k=0}^{N-1}(PS_A(k))^j. \tag{4.31}$$

With this normalization, $M_1^A = \sum_{k=0}^{N-1} PS_A(k) = N_A N$. Our experimental results on various datasets have proved that this is a good normalization. As mentioned before, zeroth moment may not be useful since it is too large. We can improve the outcomes by considering a new j-th moment:

$$M_j^A = a_j^A \sum_{k=1}^{N-1} (PS_A(k))^j. \tag{4.32}$$

It is easy to check that

$$\sum_{k=1}^{N-1} PS_A(k) = N_A N - PS_A(0) = N_A N - N_A^2 = N_A(N - N_A). \tag{4.33}$$

Therefore, we can naturally consider

$$M_j^A = \frac{1}{N_A^{j-1}(N - N_A)^{j-1}} \sum_{k=1}^{N-1} (PS_A(k))^j. \tag{4.34}$$

The fact that higher moments tend to zero is verified as follows:

$$M_j^A = N_A(N - N_A) \sum_{k=1}^{N-1} \left(\frac{PS_A(k)}{N_A(N - N_A)}\right)^j = N_A(N - N_A) \sum_{k=1}^{N-1} z_k^j, \tag{4.35}$$

where $z_k = PS_A(k)/N_A(N - N_A)$. Notice that $\sum_{k=1}^{N-1} z_k = 1$, thus it is obvious that

$$\lim_{j \to \infty} \sum_{k=1}^{N-1} z_k^j = 0.$$

Additionally, due to the symmetric property of DFT coefficients, we only have to consider the first half of the power spectrum. Therefore, the moments are improved as follows:

$$M_j^A = \frac{1}{N_A^{j-1}(N - N_A)^{j-1}} \sum_{k=1}^{[N/2]} (PS_A(k))^j. \tag{4.36}$$

The moments for the remaining nucleotides T, C, and G are derived in a similar manner. Subsequently, the first few moments are utilized to construct vectors within the Euclidean space. Our experimental findings demonstrate that the inclusion of three moments suffices for achieving accurate clustering. As a result, each gene or genome sequence can be represented as a geometric point within a 12-dimensional Euclidean space, denoted as

$(M_1^A, M_1^T, M_1^C, M_1^G, M_2^A, M_2^T, M_2^C, M_2^G, M_3^A, M_3^T, M_3^C, M_3^G)$. To cluster the gene or genome sequences, pairwise Euclidean distances are computed between these points, thereby facilitating the clustering process.

4.4.3   Cumulative Power Spectrum Moment Method

A variant of the power spectrum moment method is the following cumulative power spectrum (CPS) moment method [53]. We consider the cumulative function of the power spectrum

$$CPS_\alpha(k) = \sum_{n=1}^{k} PS_\alpha(n), \quad k = 1, 2, \ldots, N - 1. \tag{4.37}$$

It is worth noticing that we delete $PS(0)$ similar to previous methods.

For $\alpha \in \{A, T, G, C\}$, we can repeat what we have done in the previous part and define

$$M_j^\alpha = a_j^\alpha \sum_{k=1}^{N-1} (CPS_\alpha(k))^j. \tag{4.38}$$

Since

$$\sum_{n=1}^{N-1} CPS_\alpha(n) = \sum_{n=0}^{N-1} \sum_{k=0}^{n} PS_\alpha(k) - NPS_\alpha(0)$$

$$\leq N \sum_{k=0}^{N-1} PS_\alpha(k) - NPS_\alpha(0) \tag{4.39}$$

$$= NN_\alpha(N - N_\alpha),$$

a natural idea is to let $a_j^\alpha = (\frac{1}{NN_\alpha(N-N_\alpha)})^{j-1}$. However, in this way

$$M_1^\alpha = a_1^\alpha \sum_{k=1}^{N-1} CPS_\alpha(k) = \sum_{k=1}^{N-1} CPS_\alpha(k), \tag{4.40}$$

and M_1^α will be too large since CPS_α is cumulative. Therefore, the scale factor is chosen as $\frac{1}{N(NN_\alpha(N-N_\alpha))^{j-1}} = \frac{1}{(N_\alpha(N-N_\alpha))^{j-1}N^j}$ for the CPS method. Then

$$M_j^{\alpha} = \frac{1}{(N_{\alpha}(N - N_{\alpha}))^{j-1} N^j} \sum_{k=1}^{N-1} (CPS_{\alpha}(k))^j \qquad (4.41)$$

The mean value of the CPS is defined as

$$Mean_{\alpha} = \frac{1}{N - 1} \sum_{n=1}^{N-1} CPS_{\alpha}(n). \qquad (4.42)$$

We can make use of the mean value and define the central moment vectors.

$$CM_j^{\alpha} = \frac{1}{(N_{\alpha}(N - N_{\alpha}))^{j-1} N^j} \sum_{k=1}^{N-1} |CPS_{\alpha}(k) - Mean_{\alpha}|^j. \qquad (4.43)$$

The absolute value is used, otherwise the first central moment vector would be zero.

When evaluating the moments of cumulative Fourier power spectra for genomic sequences, a notable observation arises: the moment vectors and central moment vectors of the third moment are significantly smaller in magnitude compared to the first and second moments. Building upon this insight, we opt to focus exclusively on the initial two moment vectors and the first two central moment vectors. This results in a 16-dimensional truncated moment vector for each sequence in the Euclidean space, given by

$$(M_1^A, M_2^A, CM_1^A, CM_2^A, M_1^T, M_2^T, CM_1^T, CM_2^T, M_1^C, M_2^C, CM_1^C,$$
$$CM_2^C, M_1^G, M_2^G, CM_1^G, CM_2^G).$$

We name this method the cumulative Fourier power spectrum (CPS) to distinguish it from the traditional power spectrum approach. A pivotal enhancement offered by the CPS method is that, mathematically, $CPS_{\alpha}(k)$ and its moment vectors can be computed from each other, while the power spectrum cannot achieve this [53]. (A similar proof will be given in details in Theorem 6.1. The key difference between PS and CPS is that $CPS_{\alpha}(k)$ is increasing while $PS_{\alpha}(k)$ is not.) Consequently, the CPS method retains more essential information from the original sequence during the transformation into numerical sequences. This feature distinguishes it as a potent tool for preserving the inherent characteristics of the original genomic data.

Chapter 5
Graphical Representation of Sequences and Its Application

Mathematical analysis of large-volume genomic DNA sequence data is one of the challenges for biologists. Graphical representation of DNA or protein sequences provides a simple way of viewing, sorting, and comparing sequence similarity. In this chapter, we introduce two directions to construct graphical representation for biological sequences. The first direction is by curves without degeneracy and the second one is by Chaos Game Representation.

5.1 Graphical Representation by Curves Without Degeneracy

5.1.1 A Construction Without Degeneracy

Approximately two decades ago, Hamori introduced the concept of utilizing a three-dimensional H curve to depict a DNA sequence [54]. However, generating the H curve requires sophisticated computer graphic tools. As an alternative, Gates proposed a simpler two-dimensional graphical representation [55]. Nevertheless, Gates's graphical approach exhibits substantial degeneracy. For instance, sequences, like AGTC, AGTCA, AGTCAG, and so forth, yield identical graphical representations. In mathematical terms, this degeneracy manifests as repetitive closed loops or circuits within the DNA graph. In light of these considerations, we introduce a novel two-dimensional graphical representation for DNA sequences. This representation successfully eliminates circuits and degeneracy, establishing a one-to-one correspondence between DNA sequences and DNA graphs [48].

As shown in Fig. 5.1a, we construct a pyrimidine-purine graph on two quadrants of the Cartesian coordinate system, with pyrimidines (T and C) in the first quadrant and purines (A and G) in the fourth quadrant. The unit vectors representing four nucleotides A, C, G, and T are as follows:

Fig. 5.1 The unit vectors designed by Yau (**a**) and Gates (**b**) in the Cartesian coordinate plane

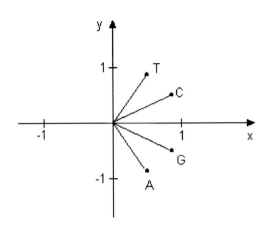

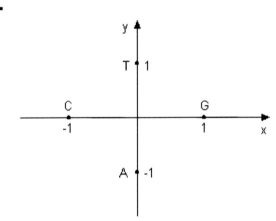

$$\left(\frac{1}{2}, -\frac{\sqrt{3}}{2}\right) \to A, \ \left(\frac{\sqrt{3}}{2}, \frac{1}{2}\right) \to C, \ \left(\frac{\sqrt{3}}{2}, -\frac{1}{2}\right) \to G, \ \left(\frac{1}{2}, \frac{\sqrt{3}}{2}\right) \to T. \qquad (5.1)$$

Different from Gates's method (Fig. 5.1b), our representation method utilizes only two quadrants of the Cartesian coordinates. The point corresponding to the jth nucleotide in the graphical representation is obtained by the sum of vectors representing nucleotides from first to jth in the sequence. Figure 5.2 illustrates the graphs for two DNA segments based on our method and Gate's method, respectively. We can see that graphs produced by our method avoid circuits.

It is not hard to prove that there is no circuit or degeneracy in our two-dimensional graphical representation. We here present a simple proof by contradiction. We assume that (1) the number of nucleotides forming a circuit is n, (2)

Fig. 5.2 Two-dimensional graphs of both human and mouse β-globin exon-1 DNA sequences are generated by Yau's (**a**) or Gates's (**b**) method. Both sequences are obtained from NCBI GenBank (AF527577 or gi:22094826 for human β-globin, and J00413 or gi:193793 for mouse β-globin)

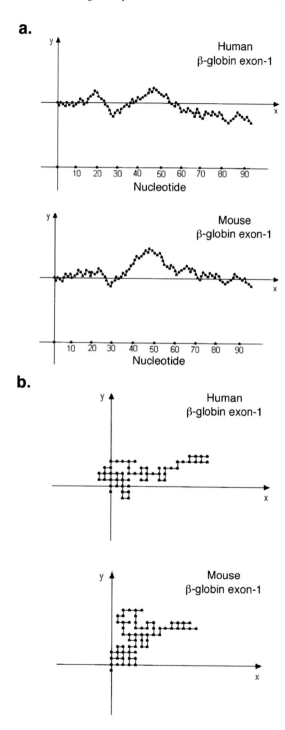

the number of A, C, G, and T in a circuit is a, c, g, and t, respectively. So, $a + c + g + t = n$. Because aA, cC, gG, and tT form a circuit, the following equation holds:

$$a\left(\frac{1}{2}, -\frac{\sqrt{3}}{2}\right) + c\left(\frac{\sqrt{3}}{2}, \frac{1}{2}\right) + g\left(\frac{\sqrt{3}}{2}, -\frac{1}{2}\right) + t\left(\frac{1}{2}, \frac{\sqrt{3}}{2}\right) = 0 \qquad (5.2)$$

then

$$a + \sqrt{3}c + \sqrt{3}g + t = 0, \quad -\sqrt{3}a + c - g + \sqrt{3}t = 0. \qquad (5.3)$$

Clearly (5.2) and (5.3) hold if and only if $a = c = g = t = 0$. Therefore, $n = 0$, which means no circuit exists in this graphical representation.

Furthermore, given x-projection and y-projection of any point $p = (x, y)$ on the sequence, we have

$$a\left(\frac{1}{2}, -\frac{\sqrt{3}}{2}\right) + c\left(\frac{\sqrt{3}}{2}, \frac{1}{2}\right) + g\left(\frac{\sqrt{3}}{2}, -\frac{1}{2}\right) + t\left(\frac{1}{2}, \frac{\sqrt{3}}{2}\right) = (x, y) \qquad (5.4)$$

and

$$a + \sqrt{3}c + \sqrt{3}g + t = 2x, \quad -\sqrt{3}a + c - g + \sqrt{3}t = 2y, \qquad (5.5)$$

where x is the x-projection and y is the y-projection of the point. $2x$ and $2y$ are irrational numbers of form $m + n\sqrt{3}$, where m and n are integers. After uniquely determining m_x, n_x, m_y, and n_y from $2x$ and $2y$, the number a_p, c_p, g_p, and t_p of A, C, G, and T from the beginning of the sequence to the point p can be found by solving the following linear system:

$$a_p + t_p = m_x$$

$$g_p + c_p = n_x$$

$$-g_p + c_p = m_y$$

$$-a_p + t_p = n_y.$$

By successive x-projection and y-projection of points on the sequence, we can recover the original DNA sequence uniquely from the DNA graph.

The current approach introduces a direct plotting method to represent DNA sequences without encountering degeneracy. In comparison to previous methods, this graphical representation is more in-line with the conventional recognition of linear sequences from the 5' to 3' end, a perspective familiar to molecular biologists. Furthermore, it can be easily constructed without the need for extensive computer graphic tools. The distinctive peaks and valleys generated by the DNA graph offer clear differentiation for specific DNA sequences, allowing for visual recognition of these long-range patterns. Through the DNA graph, one can mathematically recapture the usage of A, C, G, and T, along with the original DNA sequence, without any loss of textual information. The considerable complexity and degeneracy that have plagued previous DNA graphical representations, limiting the practical application of DNA graphs, are effectively addressed by this newly introduced two-dimensional graphical representation of DNA sequences.

5.1.2 Other Constructions Without Degeneracy

Previously, we introduced a two-dimensional graphical representation for gene sequences that eliminates circuitry and degeneracy, ensuring a one-to-one correspondence between gene sequences and gene graphs. This approach facilitates the mathematical recovery of the original DNA sequence from its graph, preserving all biological information. In this part, we will introduce two other constructions without degeneracy. They are modified versions of the method introduced in Sect. 5.1.1 and both of them help define the distance for sequences.

5.1.2.1 A Construction with Corresponding Moment Vectors

In this part, we present a slight modification to our previous method, resulting in a novel approach for graphical representation of DNA sequences [57]. The key advancement lies in our ability to derive moment vectors from DNA sequences using this novel graphical technique. The unique aspect of our approach is that these moment vectors enable the construction of a genome space within the Euclidean space. Remarkably, each genome sequence can be transformed into a distinct point within this genome space, exemplifying the distinctive feature of our method. Consequently, the genome space serves as a platform for conducting comparative analyses, facilitating investigations into clustering and phylogenetic relationships among genomes. The biological or evolutionary distance between two genomes is accurately gauged through the calculation of the Euclidean distance between the corresponding points within the genome space.

In this construction, the vectors corresponding to the four nucleotides A, C, G, and T are as follows (Fig. 5.3):

$$\left(1, -\frac{1}{3}\right) \to A, \ \left(1, \frac{2}{3}\right) \to C, \ \left(1, -\frac{2}{3}\right) \to G, \ \left(1, \frac{1}{3}\right) \to T. \tag{5.6}$$

In Fig. 5.4, we present the graphical depictions of complete mitochondrial genome sequences for the human, common chimpanzee, Norway rat, and hedgehog. These visual representations are constructed based on the vector system outlined in Fig. 5.3. Notably, the human and chimpanzee, both belonging to the primate

Fig. 5.3 Nucleotide vector
system based on $A(1, -1/3)$,
$C(1, 2/3)$, $G(1, -2/3)$ and
$T(1, 1/3)$

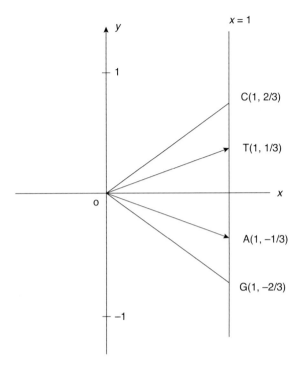

order, exhibit striking visual similarity in their mitochondrial genome graphical
representations. Because the genome lengths of these four species are similar, the
X-values of the endpoints of their graphical curves are very similar. For the Y-values
of the endpoints, the human and the common chimpanzee are very close (more than
1600), but the mouse is below 1500 and the hedgehog is below 1000.

In this construction, we can transform the graphical curves into moment vectors
motivated by Liu et al. [56]. Given the graphical curve of a DNA sequence, which
can be represented by a sequence of points $(1, y_1)$, $(2, y_2)$,...,(n, y_n), we can
consider a sequence of numbers $1 - y_1, 2 - y_2,..., n - y_n$ and define the moment
as follows:

$$M_j = \sum_{i=1}^{n} \frac{(i - y_i)^j}{n^j}, \quad j = 1, 2, \ldots, n, \qquad (5.7)$$

where n is the number of nucleotides contained in a DNA sequence. According
to this definition, each DNA sequence has an n-dimensional moment vector
$(M_1, M_2, \ldots, M_n)$ associated with it. The crucial point here is that the correspondence between a DNA sequence and its moment vector obtained from its sequence
graph is one-to-one. In other words, we can compute the values in $\{i - y_i | i = 1, \ldots, n\}$ from $(M_1, M_2, \ldots, M_n)$. (A similar proof will be given in details in

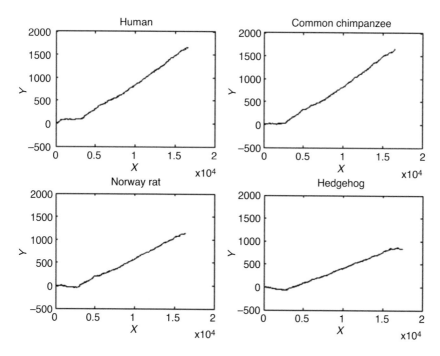

Fig. 5.4 Graphical representations of the whole mitochondrial genome sequences of four species (the human, the common chimpanzee, the Norway rat, and the hedgehog) based on the vector system shown in Fig. 5.3

Theorem 6.1. The key point is that $i - y_i$ is strictly increasing with respect to i.) Therefore, we can obtain the sequence if the moment vector is given.

In real applications, we consider the first N components $(M_1, M_2, \ldots, M_N)$ of the moment vector to represent a sequence where N is fixed for all sequences so that we can calculate the distance between two sequences with different length. Furthermore, phylogenetic and clustering analysis can be performed since the distance of two sequences is defined.

5.1.2.2   A Construction with Corresponding Probability Distribution

In this part, a new construction of graphical representation is introduced. The novelty of this method is that we can construct a probability distribution for the DNA sequences by the representation [58].

In this construction, the vectors corresponding to the four nucleotides A, C, G, and T are as follows (Fig. 5.5):

$$(1, 0.8) \rightarrow A, \ (1, 0.6) \rightarrow G, \ (1, 0.4) \rightarrow C, \ (1, 0.2) \rightarrow T. \tag{5.8}$$

Fig. 5.5 Nucleotide vector system based on $A(1, 0.8)$, $G(1, 0.6)$, $C(1, 0.4)$, and $T(1, 0.2)$

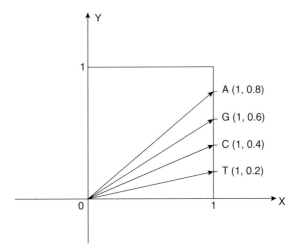

Fig. 5.6 Graphical representation of DNA sequence (ATGGTGCACC) based on the vector system shown in Fig. 5.5

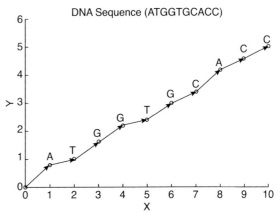

The points depicted in the graphical representation are derived by aggregating the vectors corresponding to nucleotides within the sequence. Each vector's terminal point signifies an individual nucleotide. Figure 5.6 illustrates the graphical representation of the DNA sequence (ATGGTGCACC) of the initial ten nucleotides of the human β-globin coding sequence. Importantly, the graphical curve displayed avoids circuits or redundancies, confirming a clear and unambiguous one-to-one correspondence between the sequence and its graphical representation.

Subsequently, we establish a discrete probability distribution to process the graph. In other words, we define a vector $(p_1, \ldots, p_n)$ with the following properties.

(1) $\sum_{i=1}^{n} p_i = 1$. (2) $0 \le p_i \le 1$ for $i = 1, \ldots, n$.

Our definition is shown as follows:

$$p_i = \frac{x_i - \overline{y_i}}{\frac{1}{2}n(n+1) - y_n}, \tag{5.9}$$

where (x_i, y_i) represents the position of the ith nucleotide in the DNA graphical curve and $\overline{y_i}$ represents the choice of y-coordinate value at the ith nucleotide in the DNA graphical curve according to Fig. 5.5. For example, for the DNA sequence ATGGT, we have

$$\overline{y_1} = 0.8, \quad \overline{y_2} = 0.2, \quad \overline{y_3} = 0.6, \quad \overline{y_4} = 0.6, \quad \overline{y_5} = 0.2, \quad y_5 = 2.4$$

$$(p_1, p_2, p_3, p_4, p_5) = \left(\frac{1-0.8}{\frac{1}{2}\cdot 5\cdot 6 - 2.4}, \frac{2-0.2}{\frac{1}{2}\cdot 5\cdot 6 - 2.4}, \frac{3-0.6}{\frac{1}{2}\cdot 5\cdot 6 - 2.4}, \frac{4-0.6}{\frac{1}{2}\cdot 5\cdot 6 - 2.4}, \frac{5-0.2}{\frac{1}{2}\cdot 5\cdot 6 - 2.4} \right)$$

$$= (0.0159, 0.1429, 0.1905, 0.2698, 0.3810).$$

Next, we prove that $(p_1, \ldots, p_n)$ is a discrete probability distribution:

$$\sum_{i=1}^{n} p_i = \sum_{i=1}^{n} \frac{x_i - \overline{y_i}}{\frac{1}{2}n(n+1) - y_n} = \frac{\sum_{i=1}^{n} x_i - \sum_{i=1}^{n} \overline{y_i}}{\frac{1}{2}n(n+1) - y_n} = \frac{\frac{1}{2}n(n+1) - y_n}{\frac{1}{2}n(n+1) - y_n} = 1. \tag{5.10}$$

Since $0 < \overline{y_i} < 1$ and $1 \le x_i \le n$, $x_i - \overline{y_i} \le x_i \le n$, then

$$y_n = \sum_{i=1}^{n} \overline{y_i} < n, \quad \frac{1}{2}n(n+1) - y_n > \frac{1}{2}n(n+1) - n. \tag{5.11}$$

Thus

$$p_i = \frac{x_i - \overline{y_i}}{\frac{1}{2}n(n+1) - y_n} < \frac{n}{\frac{1}{2}n(n+1) - n} = \frac{1}{\frac{n+1}{2} - 1} = \frac{2}{n-1}. \tag{5.12}$$

So, when $n \ge 3$, $p_i < 1$. On the other hand, $x_i - \overline{y_i} > 0$ and

$$\frac{1}{2}n(n+1) - y_n > \frac{1}{2}n(n+1) - n = \frac{1}{2}n(n-1) > 0, \quad n \ge 3. \tag{5.13}$$

Therefore, when $n \ge 3$, $0 < p_i < 1$. By (5.10), (5.12), and (5.13), we have proved that $(p_1, \ldots, p_n)$ is a discrete probability distribution.

The probability distribution of a DNA sequence is inherently linked to its length. It restricts the direct comparison between DNA sequences of varying lengths. To overcome this limitation, we introduce a methodology capable of generating distributions of a specified length. For a DNA sequence s with a length of n and a particular value N where $N < n$, a total of $n - N + 1$ subsequences, each of length N, can be derived. These subsequences are represented as $s_1^N, \ldots, s_{n-N+1}^N$. Each subsequence corresponds to a discrete probability distribution of length N, permitting the calculation of an average distribution across these subsequences. By adopting this approach, a probability distribution of length N can be obtained. By employing this technique, in scenarios involving a collection of sequences of varying lengths, we can standardize the process. Specifically, we can set N as the length of the shortest sequence within the dataset, thereby generating corresponding distributions of equal length for all sequences.

Now that we have discrete probability distributions with the same length N for all DNA sequences where N is the minimal length among all sequences, we want to find a dissimilarity measure between two discrete probability distributions $\lambda_1 = (p_1, p_2, \ldots, p_n)$ and $\lambda_2 = (q_1, q_2, \ldots, q_n)$. A well-known measure between two probability distributions is the Kullback-Leibler divergence [59]. The Kullback-Leibler divergence (KLD) or the relative entropy of λ_1 with respect to λ_2, denoted as $D_{KL}(\lambda_1||\lambda_2)$ is defined by

$$D_{KL}(\lambda_1||\lambda_2) = \sum_{i=1}^{n} p_i log \frac{p_i}{q_i}. \tag{5.14}$$

Kullback-Leibler divergence is often called a distance, but it is not a true distance in mathematics for not being symmetric, i.e., $D_{KL}(\lambda_1||\lambda_2) \neq D_{KL}(\lambda_2||\lambda_1)$. Moreover, it does not satisfy the triangle inequality. To make it symmetric, we use the following version:

$$d(\lambda_1, \lambda_2) := \frac{1}{2}(D_{KL}(\lambda_1||\lambda_2) + D_{KL}(\lambda_2||\lambda_1)). \tag{5.15}$$

We then utilize this method to perform phylogenetic analysis. We consider complete coding sequence of β-globin genes extracted from 10 distinct species: human (U01317), woolly monkey (AY279114), tufted monkey (AY279115), rat (X06701), rabbit (V00882), hare (Y00347), gallus (NM_001081704), duck (X15739), opossum (J03642), and salmon (NM_001123672). All these DNA sequences encompass 444 nucleotides. We utilize the UPGMA (Unweighted Pair Group Method with Arithmetic Mean) algorithm [60] to construct the phylogenetic tree, which is the diagram with branches that represents the evolutionary relationships among different biological species. The tree is depicted in Fig. 5.7. Notably, it is important to acknowledge that the phylogenetic relationships among these 10 species may not be perfectly precise due to the exclusion of complete genome information in the tree's construction. Nevertheless, the figure still illustrates the striking similarity shared by these 10 DNA sequences.

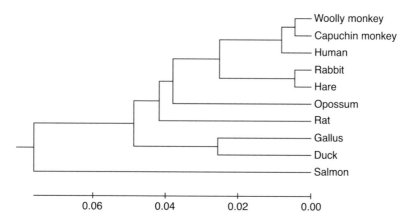

Fig. 5.7 Phylogenetic tree of 10 different species based on their complete coding sequence of β-globin genes by using our new approach

5.1.3 Constructions for Proteins

5.1.3.1 A Protein Map Based on Amino Acid Hydrophobicity

The preceding sections have introduced graphical representation methods for gene sequences. In this section, we will introduce a technique for constructing a graph from a protein sequence. Unlike gene or DNA sequences, which consist of only four nucleotides, protein sequences consist of 20 amino acids, which are more intricate. However, we demonstrate that it is possible to graphically represent protein or amino acid sequences, leading to the generation of a comprehensive protein map [61]. This protein map holds the potential to predict the properties of proteins whose functions remain undetermined.

The framework is what we introduce in Sect. 5.1.2.1. A amino acid x is mapped to a vector $(1, f(x))$ where $f(x) \in [-1, 1]$. However, since there are 20 amino acids, it is much more difficult to design their corresponding vectors. In this part, we will provide a protein map based on hydrophobicity.

Amino acid hydrophobicity is a significant property for amino acids and it plays an important role in protein folding [62]. In the protein map, we sort the amino acids based on their hydrophobicity scale values. Higher hydrophobicity scale values correspond to higher y-coordinate values. Among the amino acids, there exist 12 with positive hydrophobicity scale values. These are allocated to the first quadrant, where the difference in y-coordinate values between consecutive amino acids amounts to 1/13. Glycine, possessing a hydrophobicity scale value of zero, is assigned a y-coordinate value of zero. The remaining seven amino acids, characterized by negative hydrophobicity scale values, are positioned in the fourth quadrant. Here, the difference in y-coordinate values between neighboring amino

Table 5.1 Hydrophobicity scale values and *y*-coordinate of the 20 amino acids

Amino acid	Hydrophobicity scale	*y*-coordinate
Trp (W)	+2.25	12/13
Ile (I)	+1.80	11/13
Phe (F)	+1.79	10/13
Leu (L)	+1.70	9/13
Cys (C)	+1.54	8/13
Met (M)	+1.23	7/13
Val (V)	+1.22	6/13
Tyr (Y)	+0.96	5/13
Pro (P)	+0.72	4/13
Ala (A)	+0.31	3/13
Thr (T)	+0.26	2/13
His (H)	+0.13	1/13
Gly (G)	0	0
Ser (S)	−0.04	−1/8
Gln (Q)	−0.22	−2/8
Asn (N)	−0.60	−3/8
Glu (E)	−0.64	−4/8
Asp (D)	−0.77	−5/8
Lys (K)	−0.99	−6/8
Arg (R)	−1.01	−7/8

acids is set at 1/8. The *y*-coordinates corresponding to the 20 amino acids are enumerated in Table 5.1, and their respective vectors are illustrated in Fig. 5.8.

The points within the graphical depiction are derived from the summation of vectors that represent amino acids present in the sequence. Illustrated in Fig. 5.9, we showcase the graphical representation of the initial 10 characters of the human *β*-globin amino acid sequence. The comprehensive graphical representation of the complete human *β*-globin amino acid sequence is displayed in Fig. 5.10. A straightforward validation confirms the correspondence between the sequence and the graphical curve, ensuring a one-to-one mapping, while simultaneously ensuring that the representation is devoid of any circuits or degeneracy.

Similar to the method in Sect. 5.1.2.1, we can define the moment vector. Given the graphical curve of a DNA sequence represented by a sequence of points $(1, y_1)$, $(2, y_2), \ldots, (n, y_n)$, the moment is defined as follows:

$$M_j = \sum_{i=1}^{n} \frac{(i - y_i)^j}{n^j}, \quad j = 1, 2, \ldots, n, \tag{5.16}$$

where n is the number of amino acids contained in a protein sequence. The properties and applications of the moment vectors are almost the same as those in Sect. 5.1.2.1. We do not repeat the process.

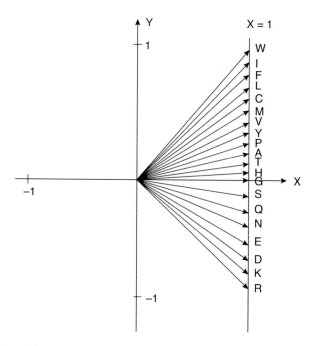

Fig. 5.8 Amino acid vector system based on Table 5.1

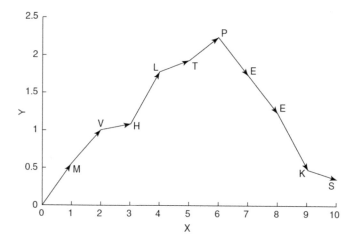

Fig. 5.9 Graphical representation of the first 10 amino acids of the human β-globin sequence based on the vector system of Fig. 5.8

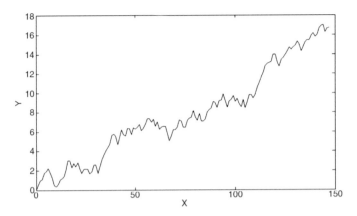

Fig. 5.10 Graphical representation of the human β-globin amino acid sequence based on the vector system of Fig. 5.8

Likewise, in practical applications, the moments are often truncated, allowing a protein sequence graph to be succinctly represented as a point situated within a two-dimensional or three-dimensional space, depending on whether the first two or the first three moments of this sequence graph are employed as its moment vector. By applying clustering techniques to either of these mappings, diverse aspects of protein sequences can be analyzed systematically and efficiently.

5.1.3.2 Protein Maps Based on Various Properties of Amino Acids

In the preceding section, the construction process of a protein map was determined by the hydrophobicity property of amino acids, overlooking other crucial biological factors governing amino acid substitution. In this section, we will present an enhanced protein map that incorporates a broader spectrum of biological elements that contribute to protein evolution at the amino acid level [63].

The influence of amino acid physico-chemical properties on substitution rates and the overall pattern of protein evolution has been extensively studied, necessitating the incorporation of a broader range of amino acid properties into our analysis [64]. In the previous research, 10 amino acid properties, including the chemical composition of side chains, polarity metrics, hydropathy, isoelectric point, volume, aromaticity, aliphaticity, hydrogenation, and hydroxythiolation, are studied [64]. Furthermore, the research conducted a comprehensive assessment of the relative significance of these amino acid properties in relation to (1) the evolution of the genetic code, (2) the amino acid composition of proteins, and (3) the pattern of nonsynonymous substitutions. The insights garnered from this investigation serve as a solid foundation for the development of our new protein map.

We formulate 10 tables to showcase the point values and corresponding y-coordinates for all amino acids, each based on one of the 10 properties under study [64, 65]. These properties include: chemical composition of the side chain, polarity, volume, polar requirement, hydropathy, isoelectric point, PC I (aliphaticity), PC II (hydrogenation), PC III (aromaticity), and PC IV (hydroxyethylation). Below, we present the isoelectric point values and their respective y-coordinates for the 20 amino acids in Table 5.2. For the amino acid with the largest property value, its y-coordinate is assigned as 1, while the amino acid with the smallest property value receives a y-coordinate of -1. Notably, the y-coordinate values for other amino acids lie within the range of -1 to 1. The discrepancy in y-coordinates between any two amino acids is proportional to the disparity in their corresponding property values. To clarify, if the value in the second column is denoted as x_i and the y-coordinate value as y_i for the ith amino acid, then the relationship holds:

$$y_i = -1 + 2 \frac{x_i - \min_j x_j}{\max_j x_j - \min_j x_j}. \tag{5.17}$$

It is important to note that several amino acids might share identical values for certain properties. For example, Leu (L), Ala (A), Val (V), Ile (I), Phe (F), and Met (M) have the same value zero with the property "chemical composition of the side chain." To address this scenario, we introduce slight perturbations to their y-coordinate values. Specifically, Phe (F) is assigned 0.001, Ile (I) receives 0.002, Val (V) is assigned 0.003, Ala (A) receives 0.004, and Leu (L) is given 0.005. This adjustment ensures that the y-coordinate values for the 20 amino acid vectors remain distinct. This distinction is pivotal to preserving the unique correspondence of moment vectors.

For one amino acid sequence, we can obtain ten graphical representations based on 10 properties. For each graphical representation, we can get one moment vector according to our protein map method. Thus, for one protein sequence with n amino acids, we have 10 moment vectors associated with it:

$$(M_{1,1}, M_{1,2}, \ldots, M_{1,n}), (M_{2,1}, M_{2,2}, \ldots, M_{2,n}), \ldots, (M_{10,1}, M_{10,2}, \ldots, M_{10,n}). \tag{5.18}$$

Since the first several moments of one n-dimensional moment vector are crucial as we discussed before, we choose the first 3 moments for each vector. Therefore, for one amino acid sequence, we have a 30-dimensional combined vector associated with it:

$$(M_{1,1}, M_{1,2}, M_{1,3}, M_{2,1}, M_{2,2}, M_{2,3}, \ldots, M_{10,1}, M_{10,2}, M_{10,3}). \tag{5.19}$$

Until this point, we have combined ten distinct physico-chemical properties of amino acids. However, the importance among these properties can exhibit considerable variability [64]. Consequently, it becomes imperative to introduce weighting factors to the constituents of the 30-dimensional vectors based on their relative

Table 5.2 Isoelectric point values and y-coordinate values of 20 amino acids

Amino acid	Isoelectric point	y-coordinate
Arg (R)	10.76	1
Lys (K)	9.74	0.74468
His (H)	7.59	0.20651
Pro (P)	6.30	−0.1164
Thr (T)	6.16	−0.15144
Ile (I)	6.02	−0.18648
Ala (A)	6.00	−0.19149
Leu (L)	5.98	−0.1965
Gly (G)	5.97	−0.199
Val (V)	5.96	−0.2015
Trp (W)	5.89	−0.21902
Met (M)	5.74	−0.25657
Ser (S)	5.68	−0.27159
Tyr (Y)	5.66	−0.2766
Gln (Q)	5.65	−0.2791
Phe (F)	5.48	−0.32165
Asn (N)	5.41	−0.33917
Cys (C)	5.07	−0.42428
Glu (E)	3.22	−0.88736
Asp (D)	2.77	−1

significance. In their comprehensive study of ten protein-coding mitochondrial genes from 19 mammalian species, Xia and Li [64] identify intriguing trends. They observe that the genetic code seems to have evolved toward minimizing polarity and hydropathy, while the other properties appear less influential. Furthermore, only the chemical composition and isoelectric point seem to have impacted the amino acid composition of the proteins under scrutiny. Regarding amino acid nonsynonymous substitutions, all ten properties, with the exception of PC IV, exert an effect. The authors provide quantified insights into the effects of these amino acid properties on the rates of substitution, measured in numerical values. Each value represents the average percentage change across the 10 genes for a specific amino acid property. The magnitude of this percentage denotes the relative importance of the property. These values serve as the weights that empower the enhancement of the 30-dimensional vector. Despite Xia and Li's assertion that PC IV has negligible influence on amino acid substitution rates [64], we assign it a smallest weight (0.1). As a result, for any given protein sequence, we derive a weighted 30-dimensional vector, delineated as follows:

$$(0.2909 M_{1,1}, 0.2909 M_{1,2}, 0.2909 M_{1,3}, 0.3240 M_{2,1}, 0.3240 M_{2,2}, 0.3240 M_{2,3},$$
$$0.2990 M_{3,1}, 0.2990 M_{3,2}, 0.2990 M_{3,3}, 0.3749 M_{4,1}, 0.3749 M_{4,2}, 0.3749 M_{4,3},$$
$$0.2358 M_{5,1}, 0.2358 M_{5,2}, 0.2358 M_{5,3}, 0.4348 M_{6,1}, 0.4348 M_{6,2}, 0.4348 M_{6,3},$$
$$0.2238 M_{7,1}, 0.2238 M_{7,2}, 0.2238 M_{7,3}, 0.1736 M_{8,1}, 0.1736 M_{8,2}, 0.1736 M_{8,3},$$
$$0.2819 M_{9,1}, 0.2819 M_{9,2}, 0.2819 M_{9,3}, 0.1 M_{10,1}, 0.1 M_{10,2}, 0.1 M_{10,3}).$$

$$(5.20)$$

By the method mentioned above, we transform the protein sequences into 30-dimensional vectors based on various properties. This vector contains more information than that based on only amino acid hydrophobicity.

5.1.4 Yau-Hausdorff Distance

In the previous parts, we have introduced some constructions of graphical representation by curves that have corresponding vectors or distributions which allow us to calculate distance. However, it is also a good idea to define a distance for curves directly. In this part, we will introduce Yau-Hausdorff distance [66], a method to measure the distance between sets of high-dimensional points. We can apply it to the curves in graphical representation since these curves can be regarded as sets of points.

We first introduce Hausdorff distance, one of the most widely used criteria for point set comparisons [67]. For two point sets A and B, the Hausdorff distance between point A and point B sets is defined by

$$h(A, B) = \max\{\max_{a \in A} \min_{b \in B} |a - b|, \max_{b \in B} \min_{a \in A} |b - a|\}. \tag{5.21}$$

When comparing graphical representations of DNA or protein sequences, our primary focus lies in the extent of shape similarity. Consequently, the ideal metric should encompass the optimal alignment considering rigid transformations, including translation and rotation. While the general Hausdorff distance gauges the separation between two fixed sets, it may not be the most suitable choice for an ideal metric, despite being a formally defined metric.

The introduced minimum two-dimensional Hausdorff distance, outlined below, serves as a well-defined metric that precisely fulfills this requirement.

$$H^2(A, B) = \min_{\theta \in [0, 2\pi]} \min_{t \in R^2} h(A + t, B^\theta), \tag{5.22}$$

where h is the Hausdorff distance defined in Eq. (5.21), and $h(A + t, B^\theta)$ stands for the Hausdorff distance between A and B after shifting A rightward by t and rotating B counterclockwise by θ.

The minimum two-dimensional Hausdorff distance is widely utilized for graph comparison, and several algorithms have been developed to calculate this distance.

Some of these algorithms assume that the point sets consist only grid points. Such algorithms are primarily employed in pixel image matching tasks, such as photo identification and MRI analysis. However, when comparing graphical representations of sequences, precise rotation of each point is required, rendering the grid point assumption inadequate.

In more general scenarios, where two point sets of sizes m and n are considered, the Huttenlocher algorithm computes the minimum two-dimensional Hausdorff distance with a time complexity of $O((m+n)^6 \log(mn))$ [68]. A later improvement to the algorithm brings down the complexity to $O((m+n)^5 \log^2(mn))$ [69]. Despite these enhancements, these algorithms remain impractical for comparing graphic curves of lengthy sequences exceeding 10,000 base pairs.

If the points are in $\mathbb{R}$, the computation will be much easier since there will be no rotation. The minimum one-dimensional Hausdorff distance [70] is defined below:

$$H^1(A, B) = \min_{t \in R} h(A + t, B), \tag{5.23}$$

where $h(A + t, B)$ is the Hausdorff distance between A and B after shifting A rightward by t. This equation can be rewritten as

$$H^1(A, B) = \min_{t \in R} \max\{\max_{a \in A+t} \min_{b \in B} |a - b|, \max_{b \in B} \min_{a \in A+t} |b - a|\}. \tag{5.24}$$

Li has proposed an algorithm to calculate the minimum one-dimensional Hausdorff distance with complexity $O((m+n) \log(m+n))$ [71].

The Yau-Hausdorff distance is a method to deal with two-dimensional points by the minimum one-dimensional Hausdorff distance:

$$D(A, B) = \max\{\max_{\theta} \min_{\varphi} H^1(P_x(A^{\theta}), P_x(B^{\varphi})), \max_{\varphi} \min_{\theta} H^1(P_x(A^{\theta}), P_x(B^{\varphi}))\}, \tag{5.25}$$

where $P_x(A^{\theta})$ is a one-dimensional point set representing the projection of A on the x-axis after being rotated counterclockwise by θ.

The Yau-Hausdorff distance D possesses the following properties [66]:

- D can be proven as a metric.
- D is defined in terms of and inherits properties from the minimum one-dimensional Hausdorff distance.
- Using the projection of two-dimensional point sets, D successfully avoids calculation of the Hausdorff distance of two-dimensional sets and can be computed efficiently.

The Yau-Hausdorff distance is not equal to the two-dimensional minimum Hausdorff distances. In fact, the Yau-Hausdorf distance is the lower bound of the minimum two-dimensional Hausdorff distances, i.e., $H^2(A, B) \geq D(A, B)$.

When dealing with two sequences of lengths m and n, the computational complexity of the Yau-Hausdorff distance between these sequence curves is $O(mn)$

Fig. 5.11 Phylogenetic tree
of COI sequences

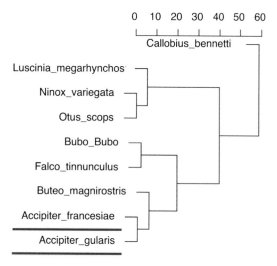

times the minimum one-dimensional Hausdorff distance. As a result, the complexity of our algorithm amounts to $O(mn(m + n)\log(m + n))$. In contrast, existing algorithms aimed at determining the minimum Hausdorff distance for point sets under Euclidean motion are considerably more intricate, with the lowest complexity of $O((m + n)^5 \log^2(mn))$ [69]. The Yau-Hausdorff distance approach significantly reduces the computational complexity involved in identifying the minimum Hausdorff distance between two-dimensional shapes.

Using the Yau-Hausdorff distance, we can define the distance between the graphical representations by curves since a curve can be seen as a set containing $n+1$ points for a sequence of length n. By this method, all constructions of graphical representations can induce a distance.

We will take the construction in Sect. 5.1.1 in the following example. The UPGMA method [60] is employed to construct the phylogenetic tree using the distance matrix generated by the Yau-Hausdorff distance. (See Fig. 5.11.) This illustration involves the COI dataset, encompassing a spider and eight raptors. (It is suggested that the mitochondrial gene cytochrome c oxidase I (COI) can function as the foundation for a global bioidentification system for animals [72].) The average length of COI gene sequences is approximately 700 base pairs.

Callobius bennetti, a spider, is distinct from the group of eight raptors. The results displayed in Fig. 5.11 underscore the success of our method in accurately clustering Callobius bennetti away from the raptors. Notably, it also positions Accipiter francesiae and Accipiter gularis in closer proximity compared to other species. This arrangement aligns with biological classification, as Accipiter francesiae and Accipiter gularis both belong to the same genus, Accipiter.

The Yau-Hausdorff distance is proposed to study the sequence comparison. It can also be generalized to measure the dissimilarity between three-dimensional structures like protein structures. The definition of it is the same as the cases in the two-dimensional space. The only difference is the dimensions of point sets A and B increase to three. The 3D Yau-Hausdorff distance does not require the compared proteins to be aligned before calculation. It can measure the similarity/dissimilarity of protein structures without superimposing them together. The 3D Yau-Hausdorff is a natural generalization for the minimum one-dimensional Hausdorff distance and takes all possible translation and rotation into full consideration. The complexity of this new 3D metric is the same as that of the two-dimensional Yau-Hausdorff distance. It is lower than many other comparison algorithms by descending dimension in the calculation without losing information of structure. These advantages enable it to be a powerful tool for comparing protein structures.

5.2 Chaos Game Representation

5.2.1 Chaos Game Representation for DNA Sequences

Chaos Game Representation (CGR) for DNA sequences is an iterative mapping technique that assigns each nucleotide in a DNA sequence to a respective position in the unit square $[0, 1] \times [0, 1]$ [73]. CGR produces a one-to-one correspondence between sequences and graphs. Given the point of ith nucleotide in the CGR, we can reconstruct the sequence from position 1 to position i. Therefore, CGR is not only a nucleotide mapping but also a sequence mapping. Another advantage of CGR is that images obtained from parts of a genome present a similar structure as that of the whole genome, which allows us to compare the genomes that are not complete.

Given the unit square in the Euclidean plane, four vertices in the unit square are assigned to the four nucleotides as $A = (0, 0)$, $T = (1, 0)$, $C = (0, 1)$ and $G = (1, 1)$. Starting with the center of the square $(\frac{1}{2}, \frac{1}{2})$, the CGR position of each nucleotide of the DNA sequence is calculated by moving a pointer to half the distance between the previous point and the corner square of the current nucleotide. The formal definition of CGR is given by an iterated function as in the equations below. For a DNA sequence $s_1 s_2 \ldots s_n \ldots s_N$, the corresponding CGR sequence $X_n = (x_n, y_n)$ is given by:

$$X_0 = (\frac{1}{2}, \frac{1}{2}), \ X_n = \frac{1}{2}(X_{n-1} + W_n), \tag{5.26}$$

where W_n is coordinates of the corners of the unit square $A = (0, 0)$, $T = (1, 0)$, $C = (0, 1)$, and $G = (1, 1)$ if s_n is A, T, C, G, respectively.

The unit square can be equally divided into four quadrants where points in the same quadrant have the same current nucleotide. Similarly, we can equally divide the square into 4^k pieces and points in each piece have the same k-mer string before the current stage. For example, we have $s_n = A$ if $X_n \in [0, \frac{1}{2}] \times [0, \frac{1}{2}]$ and we have $s_n = s_{n-1} = A$ if $X_n \in [0, \frac{1}{4}] \times [0, \frac{1}{4}]$. Therefore, each part of the CGR image has a direct biological meaning. It is not hard to see that subsequences of a gene or

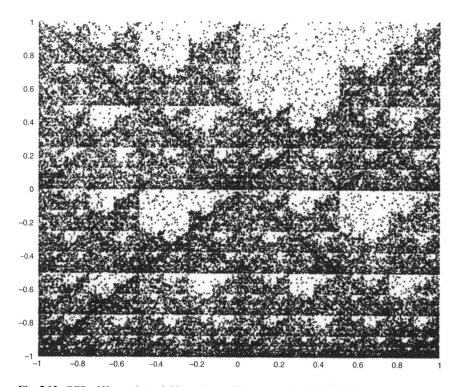

Fig. 5.12 CGR of Human beta globin region on Chromosome 11 (HUMHBB)

genome exhibit the main characteristics of the whole sequence since the piece that a point belongs to is determined mainly by its local information, thus it is useful to detect the features of the genome when only part of it is available. Figure 5.12 shows the CGR image of Human beta globin region on Chromosome 11.

The result of the CGR can be seen as a sequence of two-dimensional vectors or complex numbers. We can apply many mathematical or statistical techniques to further process the CGR result.

We can use DFT to transform the CGR result into the frequency domain and then use the even scaling method in Sect. 4.4.1 to scale all sequences to the maximum length of the data set [50]. Then distance can be defined naturally in Euclidean space and the phylogenetic tree representing evolutionary relations between the DNA sequences can be constructed.

We have applied this method to the analysis of HRV (human rhinovirus) genomes, a virus associated with upper and lower respiratory diseases and a predominant cause of common cold and cold-like illnesses. These HRV genomes are categorized into three distinct genetic groups within the genus Enterovirus and the family Picornaviridae. Our dataset comprises three main groups: HRV-A, HRV-B, and HRV-C, totaling 113 genomes, along with three additional outgroup sequences of HEV-C. Although previous studies have accurately classified the genomes, the computational time required was long due to the utilization of multiple sequence alignment for constructing the evolutionary tree.

In this study, we apply the method above (CGR+DFT) to calculate the distance among HRV genomes and utilize UPGMA [60] to draw the phylogenetic tree. The resulting phylogenetic tree, displayed in Fig. 5.13, clearly delineates the three HRV groups and effectively distinguishes them from the outgroup HEV-C, all achieved in a mere 7 seconds of computation. To provide a comparison, we also utilized Clustal Omega for clustering the dataset [52]. While Clustal Omega successfully classified the genomes into the correct groups, the process consumed 19 minutes and 35 seconds to complete. (Computations are done on a PC with configuration of Intel Core i7 CPU 2.40 GHz and 8 Gb RAM.)

The CGR result can also be used as feature inputs for machine learning. For example, the artificial neural network can be applied to detect the acceptor and donor splice sites using CGR results. Computational experiments indicate that this approach gives good accuracy [74].

5.2.2 Chaos Game Representation for Proteins

In this part, we will introduce the Chaos Game Representation for proteins. It is much more complicated than that of DNA since the number of amino acids is much higher than that of nucleobases. A natural idea is to distribute 20 amino acids on the vertices of a regular 20-sided polygon [75]. However, this method cannot be used to demonstrate the similarity of homologous protein sequences with conservative substitutions. Another method first divides amino acids into 4 types based on the polarity and then uses each vertex of the unit square to represent one type similar to the mapping in CGR for DNA [76]. However, the correspondence between sequences and representations is no longer one-to-one.

In the following content, we will introduce a three-dimensional Chaos Game Representation for proteins that maps sequences into a regular dodecahedron [77]. The regular dodecahedron is one of the five regular polyhedrons in $\mathbb{R}^3$ and its vertex number is 20. We distribute 20 amino acids on these vertices such that similar amino acids have shorter distances in the dodecahedron, which makes homologous proteins correspond to similar CGR images. The distribution arrangement is shown in Fig. 5.14. (In order to show the arrangement clearly, we unfold the dodecahedron into a planar graph.)

The dodecahedron is inscribed to ball B, which is centered at $(1,1,1)$ with radius 1. For a protein sequence $s_1 s_2 \ldots s_n \ldots s_N$, the corresponding CGR sequence $X_n = (x_n, y_n)$ is given by:

$$X_0 = (1, 1, 1), \quad X_n = (1 - u)W_n + uX_{n-1}, \tag{5.27}$$

where W_n is the coordinates of the vertex that s_n corresponds to and u is a parameter to be chosen.

Fig. 5.13 Phylogenetic trees
of 113 HRV genomes and 3
outgroup sequences. The first
graph is based on CGR and
DFT. The second graph is
based on Clustal Omega

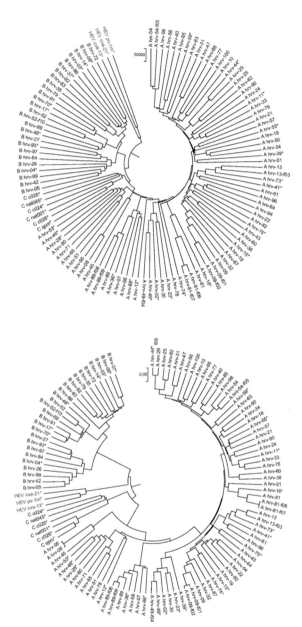

Fig. 5.14 Distribution of
amino acids on the vertices of
a regular dodecahedron
(extended image)

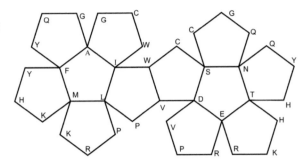

Definition 5.1 For each amino acid s whose corresponding vertex is W, we define
the ball centered at $X_s = (1-u)W + u(1, 1, 1)$ with radius u to be the ball controlled
by amino acid s, and is denoted by B_s.

Definition 5.2 For each dipeptide $s_1 s_2$ where s_1, s_2 correspond to vertices W_1, W_2,
respectively, we define the ball centered at $X_{s_1 s_2} = (1-u)W_2 + u(1-u)W_1 + u^2(1, 1, 1)$ with radius u^2 to be the ball controlled by dipeptide $s_1 s_2$ and is denoted
by $B_{s_1 s_2}$.

We can prove the following theorem:

Theorem 5.1 *For an amino acid sequence of length N, if we choose $u = \frac{\sqrt{5}-1}{\sqrt{5}+2\sqrt{3}-1}$, then the CGR images have the following properties:*

(1) $X_n \in B_{s_n}$, $\forall\, 1 \le n \le N$.
(2) $X_n \in B_{s_{n-1} s_n}$, $\forall\, 2 \le n \le N$.
(3) $B_{s_1} \cap B_{s_2} = \emptyset$ *if $s_1 \ne s_2$. More precisely, if s_1 and s_2 are adjacent vertices, then B_{s_1} and B_{s_2} are tangent to each other.*
(4) B_s *is inscribed to B.*
(5) $B_{s_1 s_2}$ *is inscribed to B_{s_2}.*
(6) *The number of amino acids s in a protein is equal to the number of points in B_s. Besides, for a certain dipeptide $s_1 s_2$, the number of $s_1 s_2$ in a protein is equal to the number of points in $B_{s_1 s_2}$.*
(7) $s_1, \dots, s_n$ *is determined by X_n, that is, given the coordinates of X_n in a CGR image, we can obtain the first n amino acids in the protein. Therefore, the three-dimensional CGR for proteins is a one-to-one correspondence to sequences.*

Proof Let s, s_1, s_2, s_n be three amino acids that correspond to W, W_1, W_2, W_n,
respectively, then

$$|X_s - X_0| = (1-u)|W - X_0| = 1 - u$$
$$|X_{s_1 s_2} - X_{s_2}| = (u - u^2)|W_1 - X_0| = u - u^2.$$

(5.28)

Fig. 5.15 A sketch map of the proof of Theorem 5.1 (3). In this figure, the vertices are denoted by ω_1, ω_2

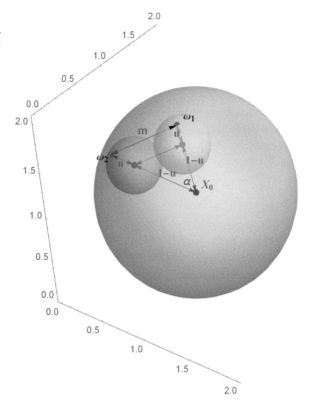

Since the radius of B, B_s, $B_{s_1 s_2}$ are $1, u, u^2$, respectively, B_s, $B_{s_1 s_2}$ are inscribed to B and B_{s_2}, which proves (4) and (5).

The mapping $f(X) = (1 - u)W_n + uX$ maps the ball B into B_{s_n} and each B_s into B_{ss_n}. It leads to (1) and (2).

The proof of (3) relies on the choice of u. Because the edge length of a regular dodecahedron inscribed to a unit ball is $m = \frac{\sqrt{3}}{3}(\sqrt{5} - 1)$, and thus the included angle of two adjacent vertices with center α satisfies $\sin \frac{\alpha}{2} = \frac{m}{2}$. According to Fig. 5.15, B_{s_1} and B_{s_2} are tangent to each other if and only if

$$\frac{u}{1 - u} = \sin \frac{\alpha}{2} \tag{5.29}$$

and therefore

$$u = \frac{\sin \alpha/2}{\sin \alpha/2 + 1} = \frac{\sqrt{5} - 1}{\sqrt{5} + 2\sqrt{3} - 1}. \tag{5.30}$$

It proves (3).

For (6), since the balls controlled by each amino acid do not intersect with each other, and X_n will be in the ball controlled by s_n, thus the number of s in a protein is equal to the number of points in B_s. Moreover, imitating the calculation in (3), we can prove that the balls controlled by different dipeptides will not intersect with each other as well. Therefore, the number of a dipeptide in a protein is equal to the number of points in the ball it controls.

For the nth point X_n in a CGR image, there exist a unique s such that $X_n \in B_s$. Therefore, the nth amino acid in the protein is s (corresponding to the vertex W) and with the formula $X_n = (1 - u)W + uX_{n-1}$, we can determine the location of X_{n-1} and the amino acid s_{n-1}. Recursively, we can determine the first n amino acids in a protein.

Therefore, we take $u = \dfrac{\sqrt{5}-1}{\sqrt{5}+2\sqrt{3}-1}$ in this method. This method can transform a protein sequence into a sequence of points in $\mathbb{R}^3$. We can use the same way as CGR for DNA to process the CGR result.

There are also some techniques established especially for the CGR result for proteins. For example, motivated by the natural vector method which will be introduced in the next chapter, we can focus on the points in each ball controlled by dipeptides (or in each group of balls) and use their statistical information to form vectors of the same length, which are practical for protein classification [77].

Chapter 6
The Development and Applications of the Natural Vector Method

In this chapter, we will introduce an effective alignment-free method called the natural vector method that can map biological sequences to vectors and discuss its properties and advantages. Then we will apply this method together to several biological problems. Finally, we show that the natural vector method can be generalized to produce some other alignment-free methods.

6.1 The Natural Vector Method for DNA Sequences

The natural vector method uses moments of nucleotides' distributions to represent the biological information in the sequence [78]. To be more specific, we combine the number, the average position, and the higher moments of each nucleotide to form a vector to represent the DNA sequence. Let n_k denote the number of the nucleotide k in the DNA sequence and let n be the length of the DNA sequence. $s[k][i]$ is the position of the ith nucleotide k in the DNA sequence. Then the average position and the higher moments of nucleotide k can be defined as follows:

$$\mu_k = \frac{\sum_{i=1}^{n_k} s[k][i]}{n_k},$$

$$D_j^k = \sum_{i=1}^{n_k} \frac{(s[k][i] - \mu_k)^j}{n_k^{j-1} n^{j-1}}, \qquad j = 2, 3, \ldots, \tag{6.1}$$

where $k = A, C, G, T$. The order m natural vector is defined as

$$(n_A, n_C, n_G, n_T, \mu_A, \mu_C, \mu_G, \mu_T, D_2^A, D_2^C, D_2^G, D_2^T, \ldots, D_m^A, D_m^C, D_m^G, D_m^T). \tag{6.2}$$

© The Author(s), under exclusive license to Springer Nature Switzerland AG 2023
S. S.-T. Yau et al., *Mathematical Principles in Bioinformatics*, Interdisciplinary
Applied Mathematics 58, https://doi.org/10.1007/978-3-031-48295-3_6

In real applications, $n_k > 0$ for each k in most cases so μ_k and D_j^k are well-defined. However, in order to make our definition more rigorous, we further define $\mu_k = D_2^k = \ldots = D_j^k = \ldots = 0$ when $n_k = 0$.

D_j^k describes the jth normalized central moment of the distribution of the nucleotide k where the normalization factor is chosen to ensure that D_j^k tends to 0 when $j \to \infty$, which will be discussed later. We start with $j = 2$ since $D_1^k = 0$.

We take a short DNA sequence $ACGGT$ as an example. We have $n_A = n_C = n_T = 1, n_G = 2$ by counting the nucleotides. The average positions are

$$\mu_A = 1,$$
$$\mu_C = 2,$$
$$\mu_G = 3.5,$$
$$\mu_T = 5,$$

and the second central moments are

$$D_2^A = 0,$$
$$D_2^C = 0,$$
$$D_2^G = \frac{(3-3.5)^2 + (4-3.5)^2}{2 \times 5} = \frac{1}{20},$$
$$D_2^T = 0.$$

Therefore, the order 2 natural vector of the sequence $ACGGT$ is

$$(1, 1, 2, 1, 1, 2, 3.5, 5, 0, 0, \frac{1}{20}, 0).$$

It is worth noticing that the natural vector method for DNA can be applied to RNA if simply replacing the base T with U. So in the following content, we do not distinguish the genetic material.

We have already obtained a good numerical characterization to represent a DNA sequence. Now we will use this tool to construct a natural vector for genomes. It is known that the structure of a genome can be very complicated. It may be single-stranded or double-stranded, and in a linear, circular, or segmented structure. Thus, we should consider the different structures when constructing the natural vector for genomes.

For the simplest genome structures, linear single-strand forms, we can treat them as linear DNA sequences. That is, every genome corresponds to a general DNA sequence. Thus, we can construct the natural vector for genomes. The order m is chosen depending on the concrete conditions. In general, $m = 2$ is enough to make the result stable, which means that we transfer a DNA sequence into a 12-

dimensional vector. Thus, using the Euclidean distance between each pair of vectors for comparison, we can perform phylogenetic and clustering analyses for genome sequences.

In the case of circular single-strand genomes, constructing the natural vector presents additional complexity due to the lack of knowledge about the starting point in the circular DNA sequence. We here propose a method to address this problem. We can consider every point as a potential starting point in the circular sequence with a length of n. This approach generates n linear single-strand genome sequences. For each linear single-strand genome, we compute its corresponding natural vector. To obtain a normalized vector, we calculate the average of these individual natural vectors. For double-stranded genomes, it is important to note that the natural vector of the reverse complementary sequence is distinct from that of the original sequence. In general, we treat double-stranded genomes as two separate single-stranded genomes. Applying the method outlined above (either linear or circular), we derive two natural vectors for these single-stranded sequences. Subsequently, we calculate the average of these two vectors to obtain a comprehensive natural vector representation. In the case of multiple-segmented sequences, we treat each segment as a sequence and transform the virus into a set of points, which will be further discussed later.

6.2   The Properties and Advantages of the Natural Vector Method

In this part, we will first introduce two important properties of natural vectors, which are the one-to-one correspondence and the convergence to 0 for high-order moments. Then we will discuss the advantages of the natural vector method.

6.2.1   The One-to-One Correspondence Between DNA Sequence and Its Natural Vector

We first show that the correspondence between a DNA sequence and its natural vector is one-to-one when the order is high enough. It is easy to see that we can calculate the natural vector given the DNA sequence. The non-trivial part of the correspondence is that we can calculate the DNA sequence given its natural vector with a sufficiently high order. To be more specific, we have the following theorem.

Theorem 6.1 *Suppose a DNA sequence contains n nucleotides where the number of four nucleotides are n_A, n_T, n_C, n_G, respectively. Then the correspondence between the DNA sequence and its order M natural vector is one-to-one where $M = \max\{n_A, n_T, n_C, n_G\}$.*

Proof In fact, we can recover the DNA sequence given the following truncation of its order M natural vector:

$$(n_A, \mu_A, n_C, \mu_C, n_G, \mu_G, n_T, \mu_T, D_2^A, \ldots, D_{n_A}^A, D_2^C, \ldots, D_{n_C}^C, D_2^G,$$
$$\ldots, D_{n_G}^G, D_2^T, \ldots, D_{n_T}^T).$$

Let $z_{[k]i} = s[k][i] - \mu_k$, then the normalized central moments can be simplified as:

$$D_j^k = \sum_{i=1}^{n_k} \frac{z_{[k]i}^j}{n_k^{j-1} n^{j-1}}, \qquad j = 2, 3, \ldots, n_k. \tag{6.3}$$

Then we have

$$\begin{cases} z_{[k]1} + z_{[k]2} + \cdots + z_{[k]n_k} = 0 \\ z_{[k]1}^2 + z_{[k]2}^2 + \cdots + z_{[k]n_k}^2 = D_2^k n_k n \\ \quad \cdots \\ z_{[k]1}^{n_k} + z_{[k]2}^{n_k} + \cdots + z_{[k]n_k}^{n_k} = D_{n_k}^k n_k^{n_k-1} n^{n_k-1}. \end{cases} \tag{6.4}$$

We take $k = A$ as an example and denote the right-hand sides by $\delta_1, \ldots, \delta_{n_A}$. Then the equations can be rewritten as

$$\begin{cases} z_1 + z_2 + \cdots + z_{n_A} = \delta_1 \\ z_1^2 + z_2^2 + \cdots + z_{n_A}^2 = \delta_2 \\ \quad \cdots \\ z_1^{n_A} + z_2^{n_A} + \cdots + z_{n_A}^{n_A} = \delta_{n_A}. \end{cases} \tag{6.5}$$

$z_1, z_2, \ldots, z_{n_A}$ are roots of a symmetric polynomial

$$(z - z_1)(z - z_2) \cdots (z - z_{n_A}) = a_0 + a_1 z + a_2 z^2 + \cdots + a_{n_A} z^{n_A}. \tag{6.6}$$

$z_1, \ldots, z_{n_A}$ can be easily solved if we can calculate $a_0, \ldots, a_{n_A}$ based on $\delta_1, \ldots, \delta_{n_A}$. The key method applied is the famous Newton's identities [80]. Let p_d ($d = 1, 2, \ldots, n_A$) be the elementary symmetric polynomials in $z_1, z_2, \ldots, z_{n_A}$, i.e.,

$$p_1 = \sum_i z_i, \, p_2 = \sum_{i<j} z_i z_j, \, p_3 = \sum_{i<j<l} z_i z_j z_l, \ldots, p_{n_A} = z_1 z_2 \cdots z_{n_A}. \tag{6.7}$$

Then $p_1 = -a_{n_A-1}, p_2 = a_{n_A-2}, \ldots, p_{n_A} = (-1)^{n_A} a_0$.
The Newton's identities claim that:

$$\delta_d - p_1\delta_{d-1} + \cdots + (-1)^{d-1}p_{d-1}\delta_1 + (-1)^d p_d = 0, \tag{6.8}$$

where $d = 1, 2, \ldots, n_A$ and p_d is the elementary symmetric polynomials in $z_1, z_2, \ldots, z_{n_A}$. Then a_i can be obtained by δ_j as shown below:

$$\begin{cases} a_{n_A} = 1 \\ a_{n_A-1} = (-1)\delta_1 \\ a_{n_A-2} = \frac{1}{2}(\delta_1^2 - \delta_2) \\ a_{n_A-3} = (-1)^3 \frac{1}{6}(\delta_1^3 - 3\delta_1\delta_2 + 2\delta_3) \\ a_{n_A-4} = \frac{1}{24}(\delta_1^4 - 6\delta_1^2\delta_2 + 3\delta_2^2 + 8\delta_1\delta_3 - 6\delta_4) \\ \cdots \quad \cdots \quad \cdots \end{cases} \tag{6.9}$$

Given the order M natural vector, $\delta_1, \ldots, \delta_{n_A}$ are known, so we can solve $a_1, \ldots, a_{n_A}$. By enumerating $l - \mu_A$ where $l = 1, \ldots, n$, we can solve all solutions of the polynomial $a_0 + a_1 z + a_2 z^2 + \cdots + a_{n_A} z^{n_A}$. Since z_i increases when i increases, we can calculate each z_i and therefore each $s[A][i]$.

Similarly, we can find all $s[k][i]$ for $k = C, G, T$, respectively. Therefore, the unique corresponding DNA sequence can be recovered based on all $s[k][i]$, $k = A, C, G, T$.

The proof above shows that the one-to-one correspondence exists rigorously when the order is sufficiently high, while in applications we can find that the one-to-one correspondence is still valid when we consider natural vectors with order 2 for most cases.

6.2.2   The Convergence to 0 for High-Order Moments

For a natural vector, the high-order moment converges to 0. The proof is as follows:

$$\begin{aligned} D_j^k &= \sum_{i=1}^{n_k} \frac{(s[k][j] - \mu_k)^j}{n_k^{j-1} n^{j-1}} \leq \sum_{i=1}^{n_k} \frac{\max_i |s[k][i] - \mu_k|^j}{n_k^{j-1} n^{j-1}} \\ &= \frac{\max_i |s[k][i] - \mu_k|^j}{n_k^{j-2} n^{j-1}} \leq \frac{n^j}{n_k^{j-2} n^{j-1}} = \frac{n}{n_k^{j-2}} \end{aligned} \tag{6.10}$$

When $n_k \geq 2$, the convergence is easy to check. If $n_k = 0$ then all moments are defined to be 0, and if $n_k = 1$ then $D_j^k = 0$. Therefore, we have proved the convergence.

From the viewpoint of probability, suppose that the expectation value of any nucleic base is $n_k = \frac{n}{4}$ (uniform distribution) for a sequence with a given length n, we can have an estimation for the upper bound:

$$\lim_j \frac{n}{n_k^{j-2}} \approx \lim_j \frac{n}{(n/4)^{j-2}} = \lim_j \frac{n \cdot 4^{j-2}}{n^{j-2}} = \lim_j \frac{4^{j-2}}{n^{j-3}}. \qquad (6.11)$$

Sometimes, n_k may be far from $\frac{n}{4}$. For example, the GC content, which is the percentage of nitrogenous bases on a DNA molecule which are either guanine or cytosine, can be extremely low for some species such as Plasmodium falciparum [79]. Even in this case, D_j^k still converges to 0 at a high speed.

The convergence for high-order moments implies that it is not necessary to involve moments of a too high order when calculating distance since high-order moments do not play important roles.

6.2.3 Advantages

There are four major advantages of the natural vector method [78].

(1) The natural vector method is much faster than alignment methods and is easier to manipulate.
(2) Once a genome space has been constructed, it can be stored in a database. There is no need to reconstruct the genome space for any subsequent application, whereas in multiple alignment methods, realignment is needed for adding new sequences.
(3) One can perform the global comparison of all genomes simultaneously.
(4) The natural vector method is not based on any invented evolutionary model, which makes the result more stable and natural.

6.3 The Natural Vector Method for Protein Sequences

We have mentioned that the natural vector method for DNA can be generalized to RNA without difficulty. Similarly, we can generalize this method to protein sequences [94]. The protein sequences are composed of 20 types of amino acids (denoted by A, R, N, D, C, E, Q, G, H, I, L, K, M, F, P, S, T, W, Y, and V, respectively). So the order m natural vector of a protein sequence can be defined by

$$(n_A, n_R, \ldots, n_V, \mu_A, \mu_R, \ldots, \mu_V, D_2^A, D_2^R, \ldots, D_2^V, \ldots, D_m^A, D_m^R, \ldots, D_m^V), \qquad (6.12)$$

which belongs to R^{20m+20}. (For most cases, $m = 2$ is good enough, leading to 60-dimensional vectors.)

The properties and advantages mentioned in the previous section are also true for this generalized version. The proof of the one-to-one correspondence and the convergence to 0 for high-order moments are the same.

6.4   The Natural Graph Method

In previous sections, we employed the phylogenetic tree as a way to depict the relationships among genomes. In this section, we will introduce the natural graph method, a more intuitive graphical representation approach that offers improved visualization of the relationships among genomes [81].

We can compute a distance matrix by calculating the Euclidean distance between every pair of natural vectors, essentially treating these vectors as points. Subsequently, we can generate a natural graph by following these steps:

(1) For each point A, find the closest point(s) B $(B_1, B_2, \ldots, B_k)$ to A. Then connect A to B $(B_1, B_2, \ldots, B_k)$ with a directed line(s) from A to B. If both A and B are closest to each other, then connect them using a bi-directional line.
(2) We then get many connected components, called level-1 graphs, after step (1). We compute the distance matrix for these connected components. The distance between two components is defined as the minimum of all distances between an element in one component and an element in another component. We then obtain a new distance matrix, in which the elements are the connected graphs obtained in step (1).
(3) Repeat the process in steps (1) and (2) to obtain higher-level graphs until we get one connected component for all elements, which is the final graphical representation.

We strengthen that other distance matrices can be used in this algorithm while we use the distance matrix produced by the natural vector method since it is a good metric for DNA or protein sequences.

Finally, we take an example to show how this method works. In this example, we only consider the closest point. Our goal is to visualize the distance matrix of 10 elements in Table 6.1. We illustrate the graph construction process in Fig. 6.1. First, we find the closest element for each element and connect them as shown in Fig. 6.1a. Then we combine the level-1 connected components to get level-2 components, graph 1 and graph 2, as shown in Fig. 6.1b. We check the minimum distance between these two graphs and get the new distance matrix in Table 6.2. The minimum distance 18 is obtained between element A in graph 1 and element G in graph 2. So, we connect these two elements to get a connected graph as shown in Fig. 6.1c. We use the directed red line to mark this connection, indicating 2nd level

Table 6.1 The distance matrix of 10 elements

	A	B	C	D	E	F	G	H	I	J
A	0									
B	9	0								
C	13	4	0							
D	23	21	23	0						
E	27	34	38	30	0					
F	26	36	39	39	12	0				
G	18	26	30	25	12	16	0			
H	19	8	9	18	34	25	25	0		
I	20	14	11	30	43	44	35	12	0	
J	28	21	20	18	20	47	37	17	20	0

connection. Clearly, this directional graphical representation uniquely illustrates the 1st-nearest-neighbor relationships.

The direction in the graph shows the closest element to each element based on their biological distances. For example, given a virus A, virologists would like to know which virus B is closest to A. An arrow from A to B in the graph represents this relation. Here we need to point out that the natural graph is not necessarily a tree. As in the example, a cycle may exist in the graphical representation which may show interesting biological information.

6.5   Applications

6.5.1   12-Dimensional Viral Genome Space

The natural vector method can help classify genomes and let us know the relationship among genomes. More specifically, this method is able to transform all genomes into points in a 12-dimensional genome space and many mathematical and statistical techniques can be applied to study these points. We take the viral genome sequences as an example to show how a 12-dimensional genome space is established. We will first introduce the genome space for only single-segmented viruses, and then we will add the multiple-segmented viruses.

6.5.1.1   The Genome Space for Only Single-Segmented Viruses

In [81], 2044 reference sequences of single-segmented viruses are considered to establish a genome space. All sequences are transformed into 12-dimensional (order 2) natural vectors. A problem we may face in real applications is that the sequencing data can be ambiguous. As is shown in Table 1.2, a single character may represent more than one nucleotide. For example, the letter R refers to either

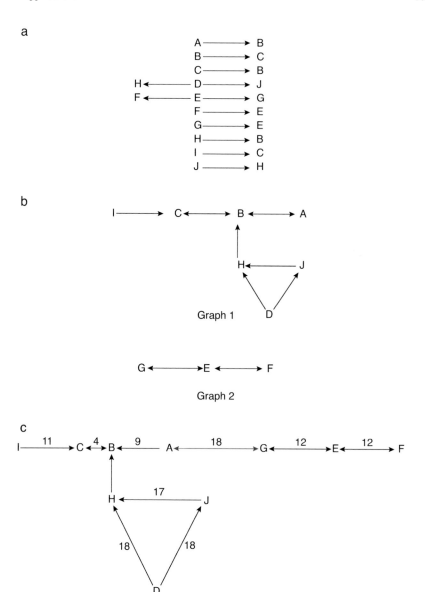

Fig. 6.1 The construction of the natural graph

A or G. In order to deal with this problem, we first introduce an equivalent definition of natural vectors. Let $S = (s_1, s_2, \ldots, s_n)$ be a nucleotide sequence of length n where $s_i \in \{A, C, G, T\}$, $i = 1, 2, \ldots, n$. For $k = A, C, G, T$, define $w_k(\cdot) : \{A, C, G, T\} \to \{0, 1\}$ such that $w_k(s) = 1$ if $s = k$ and $w_k(s) = 0$ otherwise.

Table 6.2 The distance
matrix of 2 graphs obtained
from Fig. 6.1b

	Graph 1	Graph 2
Graph 1	0	
Graph 2	18	0

(1) Let $n_k = \sum\limits_{i=1}^{n} w_k(s_i)$ denote the number of letter k in S.

(2) Let $\mu_k = \sum\limits_{i=1}^{n} i \cdot \dfrac{w_k(s_i)}{n_k}$ be the mean position of letter k.

(3) For $j = 2, 3, \ldots$, let $D_j^k = \sum\limits_{i=1}^{n} \dfrac{(i - \mu_k)^j w_k(s_i)}{n_k^{j-1} n^{j-1}}$.

The order m natural vector of a nucleotide sequence S is defined by

$$(n_A, n_C, n_G, n_T, \mu_A, \mu_C, \mu_G, \mu_T, D_2^A, D_2^C, D_2^G, D_2^T, \ldots, D_m^A, D_m^C, D_m^G, D_m^T). \tag{6.13}$$

The natural vector defined here is exactly the same as that defined previously but it is easier to be generalized when s_i is not in the range of $\{A, C, G, T\}$. To be more specific, for $k = A, C, G,$ and T, let the weight $w_k(s_i)$ be the expected count of letter k at position i. For example, R refers to either A or G, then $w_A(R) = w_G(R) = \frac{1}{2}$ and $w_C(R) = w_T(R) = 0$. As another example, N refers to $A, T, C,$ or G, then $w_A(N) = w_C(N) = w_G(N) = w_T(N) = \frac{1}{4}$. By the introduction of weight, we can deal with the problem of ambiguous sites and successfully transform viral genomes into 12-dimensional vectors.

For each virus, we can find its nearest neighbor in the 12-dimensional natural vector genome space and check whether its label matches that of its nearest neighbor. If we have a complete genome space that contains all of the viruses it is reasonable to assume any virus must have a neighbor sharing the same label. In reality, even though the genome space may be incomplete, we can still predict the label of a sequence by this strategy. There are many types of labels that can be chosen. The Baltimore class, the family, the subfamily, and the genus of a virus are examples of the label.

We first take the Baltimore class as the label. The Baltimore classification system categorizes viruses into seven classes based on the type of genome molecule and replication strategy [82–84]. These classes are double-stranded DNA viruses, single-stranded DNA viruses, double-stranded RNA viruses, positive-sense single-stranded RNA viruses, negative-sense single-stranded RNA viruses, single-stranded RNA reverse transcriptase viruses, and double-stranded DNA reverse transcriptase viruses, respectively. (Labels range from I to VII.)

With viral genome data and topological characteristics (DNA/RNA, single/double-stranded, linear/circular), natural vectors can be used to predict Baltimore class labels. Single-stranded DNA or double-stranded RNA sequences belong to class II or III, respectively. Double-stranded DNA sequences may be class I or VII, while single-stranded RNA sequences could be class IV, V, or

VI. Inconsistency, based on the percentage of viruses with inconsistent nearest neighbor labels, gauges method effectiveness. In class I and VII, linear viruses have 0% inconsistency, whereas circular viruses exhibit 1.36%. In classes IV, V, and VI, linear viruses demonstrate 6.55% inconsistency, with circular viruses at 0%. Overall, inconsistency stands at 3.18%.

Sparsity of the reference dataset leads to inconsistency, as distant neighbors become unreliable due to potential true neighbor absence. This explains inconsistency distribution. To mitigate this, we determine the 75% quantile of nearest distances within each class and set a 75%-cutoff for predictions. (We make the prediction only when the nearest distance is less than the cutoff.) This adjustment reduces circular viruses' inconsistency in classes I and VII to 0.90% and linear viruses' inconsistency in classes IV, V, and VI to 2.33%. The overall rate improves to 1.19%, prompting the adoption of this strategy for subsequent predictions.

Even with only sequence information, accurate Baltimore class predictions are possible. Among 2044 viruses, 54 receive incorrect predictions, yielding a 2.64% inconsistency rate. Higher inconsistency in classes III (11%) and VI (12%) is due to their smaller sizes (45 and 58, respectively). Expanding the database with more samples should further diminish inconsistency rates.

The classification also achieves good performance when the label is chosen as the family, the subfamily, and the genus. We can predict the family of a virus based on only sequence information with an inconsistency rate of 3.38%. The inconsistency rate of subfamily prediction and genus prediction based on sequence information and family information are 0.29 and 2.79%, respectively.

The natural graph allows us to visually represent and analyze the outcomes of the natural vector method in more detail. In Fig. 6.2, we provide a graphical depiction of 44 single-segment referenced viruses belonging to Baltimore class VII, serving as an illustrative example. Each integer in the figure corresponds to a virus, while the real number displayed on an arrow signifies the distance between two viruses.

Notably, the graph distinctly separates the Hepadnaviridae and Caulimoviridae families. Within the Hepadnaviridae family, there are two genera Avihepadnavirus and Orthohepadnavirus. For viruses No. 1476 (Ross's goose hepatitis B), No. 1529 (Sheldgoose hepatitis B), and No. 1583 (Snow goose hepatitis B) that lack ICTV genus labels, their nearest neighbors are all attributed to the Avihepadnavirus genus, leading us to predict their association with the Avihepadnavirus genus. These predictions align with the findings of other researchers [85]. The family Caulimoviridae encompasses six genera: Badnavirus, Petuvirus, Caulimovirus, Cavemovirus, Soymovirus, and Tungrovirus. Virus No. 988 (Lucky bamboo bacilliform) lacks a genus label and is nearest to Virus No. 482 (Dracaena mottle), with a distance of merely 14.52. Given this proximity, we confidently predict that Virus No. 988 is also a member of the Badnavirus genus, consistent with the conclusions of Chen et al. [86]. Furthermore, while ICTV labels Virus No. 217 (Bougainvillea spectabilis chlorotic vein-banding) and Virus No. 454 (Cycad leaf necrosis) as Badnaviruses, our genome space analysis reveals their significant separation from the rest of the Badnaviruses. Similarly, viruses such as No. 325 (Cestrum yellow leaf curling), No. 616 (Eupatorium vein clearing), and No. 1481 (Rudbeckia flower distortion) display

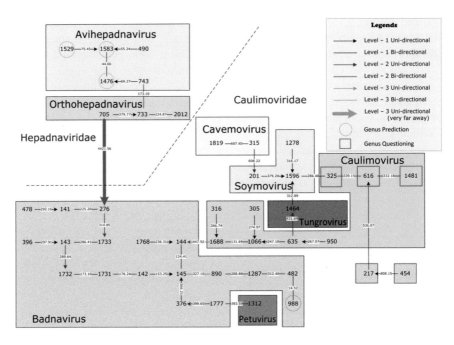

Fig. 6.2 The natural graph for the 44 single-segment referenced viruses in Baltimore class VII

considerable distance from the remaining Caulimoviruses. Consequently, we raise questions about the ICTV genus classifications for these particular viruses.

For comparison, we also conducted Multiple Sequence Alignment (MSA) analysis on three small Baltimore classes: III (45 sequences), V (67 sequences), and VII (44 sequences), aiming to assess whether a virus's nearest neighbor corresponds to the same family as itself. The ClustalW program from the MEGA 5.0 software was employed to perform the alignment for these three groups, followed by verification of label consistency between viruses and their nearest neighbors. The results indicated inconsistent family labels in Baltimore class III (1 inconsistency), while no inconsistencies were found in classes V and VII. While MSA offers accurate classification outcomes, one of its most prominent challenges is computational time. MSA demands around 2, 10, and 1 hours to generate alignment results for Baltimore classes III, V, and VII, respectively, on a standard PC (CPU 1.67 GHz, 3 GB of RAM). In contrast, our approach achieves similar results in just 2.1, 9.4, and 1.7 seconds for the same classes on the same computer. When dealing with larger classes such as Baltimore classes I (776 viruses), II (328 viruses), and IV (563 viruses), MSA becomes computationally demanding, if not impractical. In contrast, our method takes approximately 76.7 minutes, 5.2, and 45.1 seconds for these classes, with the majority of computational time allocated to natural vector calculations. It is worth noting that classifying a new virus typically requires less than one second to calculate its natural vector and determine its classification. Our approach surpasses MSA in terms of computational efficiency, as the recalculations of natural vectors are unnecessary for known viruses. For Baltimore class VI, a check is unnecessary since all 58 viruses belong to a single family.

6.5.1.2 The Genome Space with Multiple-Segmented Viruses

For multiple-segmented sequences, we transform each segment into a 12-dimensional natural vector. Thus each sequence is mapped into a set of points. To simultaneously compare viruses with multiple segments, we propose the use of the Hausdorff distance which measures the distance between two sets of vectors [87].

The definition of the Hausdorff distance has been introduced in Sect. 5.1.4. Here, we employ an example to illustrate its application for multi-segmented sequences more effectively. Let us assume that virus X is composed of four segments with corresponding natural vectors x_1, x_2, x_3, x_4, and virus Y consists of four segments with corresponding natural vectors y_1, y_2, y_3, y_4. Suppose we have the Euclidean distance matrix $(d_{ij}) = (d(x_i, y_j))$ given by:

$$\begin{pmatrix} 16 & 7 & 2 & 23 \\ 1 & 10 & 15 & 8 \\ 25 & 18 & 7 & 2 \\ 19 & 3 & 37 & 22 \end{pmatrix}. \tag{6.14}$$

We identify the smallest distance in each row (2, 1, 2, and 3, respectively) and in each column (1, 3, 2, and 2, respectively). Consequently, the Hausdorff distance between X and Y is determined as $\max(\max 2, 1, 2, 3, \max 1, 3, 2, 2) = 3$. One significant advantage of the Hausdorff distance is its invariance to the rearrangement of virus segment orders. Hence, there is no need to align the segments of two viruses beforehand for measuring their distance. Furthermore, if $h(X, Y) = 0$, it signifies that the two viruses possess identical matched segments. Another benefit lies in the Hausdorff distance's capability to compare viruses with varying segment quantities. For instance, even if segment x_4 is unintentionally omitted, we can still calculate $h(x_1, x_2, x_3, Y) = 7 > 3 = h(X, Y)$. In this scenario, the Hausdorff distance might impose a penalty due to the absent segment.

By the Hausdorff distance, we can add the multiple-segmented sequences into the genome space. It is worth noticing that we can calculate the distance between a single-segmented sequence and a multiple-segmented sequence since the Hausdorff distance does not require the numbers of the segments to be the same. Therefore, two kinds of sequences can be put into one genome space. In [87], 2384 reference sequences including 370 multiple-segmented viruses are considered to establish a genome space. Applying the Hausdorff distance, we observe that 97.7% of the 2384 viruses exhibit an identical number of segments with their closest neighbors. This suggests a tendency for a virus to have the same segment count with its nearest neighbor according to the Hausdorff distance. However, there are a few cases where a virus and its closest neighbor possess differing segment quantities. For instance, utilizing the Hausdorff distance, the nearest neighbor of the Subterranean clover stunt virus (SCSV), characterized by eight segments, is the Abaca bunchy top virus (ABTV), comprising six segments. Remarkably, both viruses belong to the Nanoviridae family.

Similar to the genome space with only single-segmented viruses, 75%-cutoff strategy is also applied when predicting Baltimore classes and other labels. The inconsistency rate for Baltimore class and family given the sequence only are 3.5% and 4.6%, respectively, and the inconsistency rate for subfamily and genus based on sequence information and the family information are 0.3% and 4.4%, respectively. The results are comparable with those in [81].

We have successfully established a genome space that encompasses multiple-segmented viruses. This genome space provides us with a valuable tool for analyzing specific viruses, such as the influenza A (H7N9) virus. The H7N9 virus, comprised of eight gene segments, poses a significant public health concern due to its high contagiousness, lethality, and rapid evolution [88]. Swift identification and placement of newly evolved H7N9 strains within the phylogenetic tree are of paramount importance. Traditionally, the phylogenetic trees for these strains are constructed on a segment-by-segment basis, often focusing solely on the hemagglutinin (HA) and neuraminidase (NA) gene segments. However, leveraging the natural vector method allows us to conduct a simultaneous analysis of these two gene segments. In our study, we consider 28 strains of H7N9 viruses sourced from the NCBI Influenza virus database. We employ the neighbor-joining method to reconstruct a phylogenetic tree using the Hausdorff distance of natural vectors (see Fig. 6.3). The clustering outcome from this approach outperforms the results obtained from the consensus tree [89], which merged the HA and NA segment trees using the majority rule. This highlights the efficacy of our method in enhancing the accuracy of phylogenetic analyses for multi-segmented viruses.

There are many other applications of the 12-dimensional viral genome space. The classification and prediction based on the natural vector method have been applied to many viruses including HIV [78], West Nile viruses [81], and Ebolavirus [90]. In order to promote the analysis based on the natural vector method, a virus database called VirusDB (http://yaulab.math.tsinghua.edu.cn/VirusDB/) and an online inquiry system has been constructed to serve people who are interested in viral classification and prediction. The database stores all viral genomes, their corresponding natural vectors, and the classification information of the single/multiple-segmented viral reference sequences downloaded from the National Center for Biotechnology Information. The online inquiry system serves the purpose of computing natural vectors and their distances based on submitted genomes, providing an online interface for accessing and using the database for viral classification and prediction, and back-end processes for automatic and manual updating of database content to synchronize with GenBank. Submitted genomes data in FASTA format will be carried out and the prediction results with 5 closest neighbors and their classifications will be returned by email. Considering the one-to-one correspondence

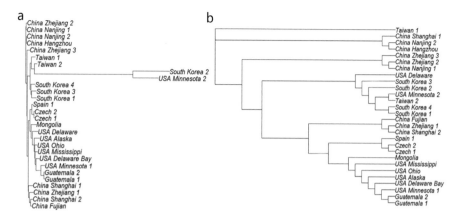

Fig. 6.3 (**a**) Phylogenetic tree based on the Hausdorff distances of the NA and HA segments. (**b**) The consensus tree of the HA and NA segments

between sequence and natural vector, time efficiency, and high accuracy, the natural vector method is a significant advance compared with alignment methods, which makes VirusDB a useful database in further research.

6.5.2 60-Dimensional Protein Space

Similarly, the natural vector method for protein sequences can be applied to show how proteins are distributed in the protein space. We take three examples to show the effectiveness of the 60-dimensional protein space constructed by the natural vector method.

6.5.2.1 The Classification of the PKC-Like Superfamily

In [94], a 60-dimensional natural vector approach is employed to classify proteins belonging to the PKC-like superfamily. Protein kinase C (PKC) is a family of enzymes that plays a pivotal role in regulating the activity of other proteins by the phosphorylation of hydroxyl groups on serine and threonine amino acid residues within these proteins [95]. The structure of PKC proteins is characterized by a regulatory domain and a catalytic domain, interconnected by a hinge region. Notably, the regulatory domain tends to determine the primary classification, given that the catalytic domain usually displays a high degree of conservation. The PKC-like superfamily is composed of six categories of PKCs and PKC-related protein molecules: cPKC, nPKC, aPKC, PKCmu (v, μ, and D2 types), PKC1 (from fungus), and PRK (similar to PKC1 but from animals).

124 proteins from the PKC-like superfamily are transformed into 60-dimensional vectors by the natural vector method. By computing the Euclidean distances between these vectors, we obtain a distance matrix. Based on the distance matrix, a natural graph can be drawn (See Figs. 6.4, 6.5, and 6.6). The lengths of lines in this graphical representation are proportional to the biological distances among the proteins. Our classifying results for these proteins totally agree with those from GenBank (NCBI) descriptions and literature and the natural graph can greatly visualize the distribution of proteins in the protein space.

Furthermore, the natural vector method provides us with insight into the relationship among proteins. For example, sequence No. 5 (GenBank ID: O17874) (in Fig. 6.6) is a protein sequence belonging to the category PRK from C. elegans. In the natural graph, it is closest to a PKC1 protein (No. 89). We further check that all the other PRK proteins in the dataset are from vertebrate animals. Thus, we believe that the PRK subfamily should be divided into two smaller groups, one is from vertebrate animals (PRK-v) and the other is from invertebrate animals (PRK-inv). Thus, sequence No. 5 belongs to a new subfamily PRK-inv, which is closer to PKC1 subfamily than PRK-v. As another example, sequence No. 84 (GenBank ID: Q69G16) (in Fig. 6.4) is a cPKC according to GenBank while the closest protein sequence from it belongs to aPKC. Thus, our theory predicts that there may be some cPKC members missing in our dataset, lying between No. 84 and No. 100 in our protein space. It is the job of biologists to find these new cPKC members. This unique natural graphical representation gives a whole picture of the phylogenetic relationships of the PKC-like superfamily. It allows us to have a global comparison of proteins simultaneously, which no other existing method can achieve.

6.5.2.2 The Evolutionary Origin of the SAR11 Clade Marine Bacteria

In [96], a 60-dimensional natural vector approach is employed to discern the phylogenetic placement of the SAR11 clade. Planktonic bacterial lineages with streamlined genomes are broadly distributed throughout the oceans. A notable example of such lineages is the SAR11 clade within the Alphaproteobacteria [97]. The SAR11 bacteria constitute a highly abundant group inhabiting the upper layers of oceanic surface waters, playing a crucial role in the global ocean carbon cycle. The Global Ocean Sampling Expedition (GOS) has confirmed SAR11's prevalence as the most dominant ribotype across various ocean habitats, encompassing coastal, estuary, and open-ocean environments [98]. However, the taxonomic delineation of the SAR11 clade within the Alphaproteobacteria remains relatively indistinct [99]. Statistical analyses often yield conflicting evolutionary models, further complicating matters. Clarifying the origin of the SAR11 lineage requires addressing the uncertainty surrounding its position in the Alphaproteobacteria tree [97]. This task presents a challenge due to the fact that the genomic G+C content of ecologically distinct SAR11 and Rickettsiales lineages remains notably low (below 30%), in contrast to the majority of other Alphaproteobacterial lineages that exhibit a higher G+C content (ranging from 50% to 70%) [97].

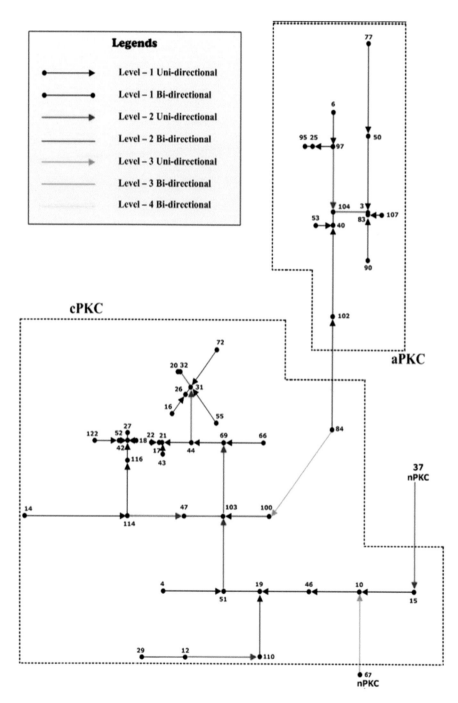

Fig. 6.4 The natural graphical representation of 124 proteins from PKC-like superfamily. We break the large original figure into three pieces: (a)–(c). This figure is Part (a)

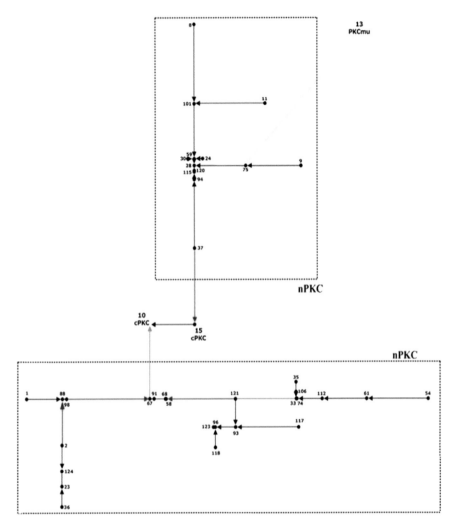

Fig. 6.5 The natural graphical representation of 124 proteins from PKC-like superfamily. This figure is Part (b)

Eight clades are used to reconstruct the phylogeny of Alphaproteobacteria and 62 strains from these clades are chosen in Luo's research [97]. (See Table 6.3.) We take the following steps to define the distance between two clades. First, we pick a protein dataset (containing some specific families) and transform each protein sequence into a 60-dimensional natural vector. Then each strain can be seen as a set of vectors whose corresponding protein sequences belong to this strain and the distance between two strains can be calculated by the Hausdorff distance. Finally, we use the Hausdorff distance again to calculate the distance between two clades by regarding each clade as a set of strains.

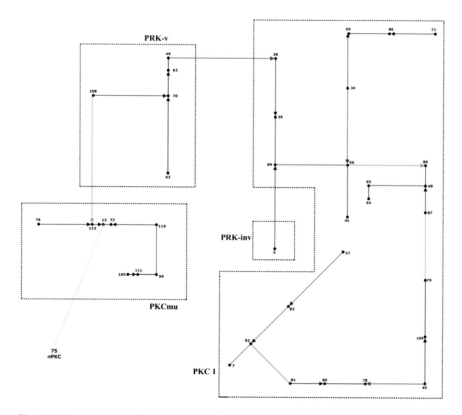

Fig. 6.6 The natural graphical representation of 124 proteins from PKC-like superfamily. This figure is Part (c)

Table 6.3 Eight clades used for reconstructing the phylogeny of Alphaproteobacteria. The number in parentheses shows the number of strains in each clade

Number	Clade name
1	Caulobacterales (5)
2	Rhizobiales (14)
3	Rhodospirillales (7)
4	Rickettsiales (7)
5	Rhodobacterales (10)
6	SAR11 (8)
7	SAR116 (5)
8	Sphingomonadales (6)

In [96], many protein datasets are used to study the evolutionary origin of the SAR11 clade. We here show the result of one dataset. The chosen protein dataset, including 3315 protein sequences, consists of 24 composition-heterogeneous ribosomal protein families and 28 composition-homogeneous protein families. Given this dataset, we can compute the distance between two clades. Then the phylogenetic tree (using the single linkage method [100] and neighbor-joining algorithm [101])

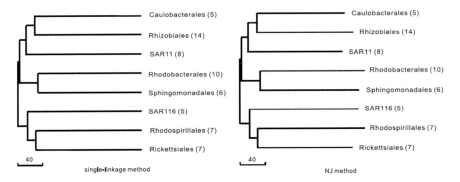

Fig. 6.7 The phylogenetic tree for Alphaproteobacteria based on the Hausdorff distance

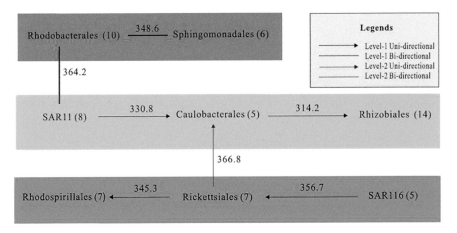

Fig. 6.8 The natural graph for Alphaproteobacteria based on the Hausdorff distance

and the natural graph can be plotted. (See Figs. 6.7 and 6.8.) The evolutionary result is consistent with Viklund's work [98].

6.5.2.3   The Phylogenetic Analysis of the Zika Virus

In [103], the phylogenetic analysis of the Zika virus (ZIKV) is conducted based on the 60-dimensional protein space. Previously we have shown the effectiveness of the genome space in the virus-related study. Now we will show that the protein space is also practical when studying viruses. Zika virus (ZIKV) is an arbovirus within the genus Flavivirus and family Flaviviridae. It is transmitted by Aedes (Stegomyia) mosquitoes and is closely related to other flaviviruses such as dengue, West Nile, and yellow fever viruses.

Most of the phylogenetic trees of the Zika virus were constructed by multiple sequence alignment algorithms and based on genomic sequences. However, these trees were inconsistent with one another. For example, the recovered evolutionary relationship presented two cases: Asian and African lineages; Asian and two African lineages. Based on the complete genomes, the two different cases of ZIKV phylogeny could be obtained [102]. The protein sequences of ZIKV are directly involved in a variety of biological processes and are more conserved than the gene sequences. To fully understand the origin and diversity of this virus, the natural vector method can be applied to perform the phylogeny analysis of ZIKV strains.

According to the organization of the ZIKV genome, the complete coding region of ZIKV is first translated into a protein (polyprotein) which is then cleaved into three structural and seven nonstructural proteins. Based on the 60-dimensional natural vector method, the phylogenetic tree using 87 ZIKV polyproteins is reconstructed. As illustrated in Fig. 6.9, the 87 ZIKV are well classified into three lineages: the Asian, West African, and East African lineages, which is consistent with the previous work [104]. The two African clades form a sister group to the Asian clade. The newly identified ZIKV strains in countries of Americas (such as Brazil, Haiti, Suriname, Guatemala, Martinique, and Puerto Rico) are all close to Asian and Pacific strains as shown in Fig. 6.9.

The strains from the Asian and Pacific countries are in the base of the Asian clade consisting of strains from Asia, Pacific regions, and Americas. This indicates that the epidemics in the Americas are likely dated back to strains from Malaysia isolated in 1966. The two African lineages are sister groups. The strains from Senegal are distributed in the two lineages, which suggests that two independent lineages have been circulating in this country. The MR766 prototype strains are clustered together as well.

6.5.2.4 The Protein Universe

The applications discussed above highlight the utility of the natural vector method in gaining insights into the distribution of proteins within the protein space. A fundamental challenge in biology revolves around unraveling the characteristics of the protein universe, encompassing the entirety of known proteins [92]. This expansive and enigmatic entity constitutes a foundational aspect of biological understanding. Prior approaches [91–93] aimed at elucidating the nature of the protein universe typically clustered sequences into families based on their similarities. However, these methods often fall short of providing a concrete depiction of the protein universe's nature. Furthermore, they rely on diverse amino acid substitution models and entail manual intervention, leading to potentially contentious outcomes. Additionally, these approaches rely on multiple sequence alignment, a computationally intensive NP-hard problem, rendering them impractical for handling substantial volumes of protein sequences. In contrast, the natural vector method for proteins offers an effective representation of a protein sequence's position within the protein universe. Notably, it presents a swift algorithm capable of parallel computation, thereby addressing the computational limitations of previous methods.

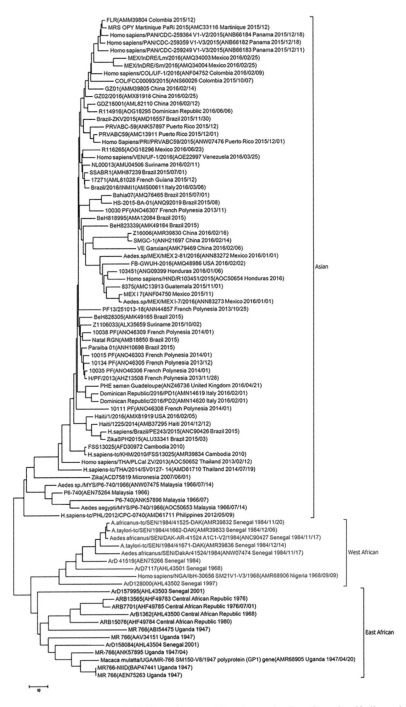

Fig. 6.9 Phylogenetic tree of 87 Zika viruses with polyproteins based on the 60-dimensional natural vector method

6.6   Other Alignment-Free Methods Motivated by the Natural Vector Method

There are many alignment-free methods that are motivated by the natural vector method. For example, in Chap. 8, we will introduce the k-mer natural vector method. In this part, we will present another generalization of the natural vector method, which takes the covariance into account.

The order 2 natural vector contains the occurrence, the mean, and the variance. In [105], the covariance is further added to the vector to produce a new alignment-free method. In the following content, we will define the covariance of two letters in a string.

We first define the covariance of two sets of points in $\mathbb{R}$. Let $A = \{a_1, a_2, \ldots, a_n\}$ and $B = \{b_1, b_2, \ldots, b_m\}$ be two finite point sets in $\mathbb{R}$, where $a_1 < a_2 < \cdots < a_n$ and $b_1 < b_2 < \cdots < b_m$.

(1) If $m = n$, then $Cov(A, B) := \sum_{i=1}^{m}(a_i - \mu_A)(b_i - \mu_B)/m$, where $\mu_A = \sum_{i=1}^{n} a_i/n$, $\mu_B = \sum_{i=1}^{m} b_i/m$.
(2) If $m \neq n$, we can assume that $n > m$. We then choose m numbers from set $A = \{a_1, a_2, \ldots, a_n\}$ which satisfy $a_{i_1} < a_{i_2} < \cdots < a_{i_m}, 1 \leq i_1 < i_2 < \ldots < i_m \leq n$, and here are C_n^m choices in total. We compute the covariance between the m numbers and set $B = \{b_1, b_2, \ldots, b_m\}$, then take the average value of these C_n^m results as the final covariance of the point sets A and B, written as I. We can simplify the result in the language of matrix, that is,

$$I = \frac{1}{mC_n^m} BDA^T - \mu_A\mu_B, \tag{6.15}$$

where $\mu_A = \sum_{i=1}^{n} a_i/n$, $\mu_B = \sum_{i=1}^{m} b_i/m$, $A^T = \{a_1, a_2, \ldots, a_n\}^T$ represents an $n \times 1$ column vector, and D is an $m \times n$ matrix written as $(D_{ij})_{m \times n}$,

If $i = 1$,
$$D_{ij} = \begin{cases} C_{n-j}^{m-1} & , 1 \leq j \leq n - m + 1 \\ 0 & , n - m + 2 \leq j \leq n \end{cases}$$

If $2 \leq i \leq m - 1$,
$$D_{ij} = \begin{cases} 0 & , 1 \leq j \leq i - 1 \\ C_{j-1}^{i-1} C_{n-j}^{m-i} & , i \leq j \leq n - m + i \\ 0 & , n - m + i + 1 \leq j \leq n \end{cases}$$

If $i = m$,
$$D_{ij} = \begin{cases} 0 & , 1 \leq j \leq i - 1 \\ C_{j-1}^{m-1} & , i \leq j \leq n \end{cases}.$$

For a DNA sequence S of length N, we want to compute the covariance between any pair of nucleotides or amino acids X and Y. Assume that position of X appeared in the sequence S is $A = \{a_1, a_2, \ldots, a_n\}$, and the position of Y is $B = \{b_1, b_2, \ldots, b_m\}$. Then the covariance between X and Y is defined as $Cov(A, B)/N$.

For example, given a DNA sequence ACACACGTGT, we first compute the covariance between nucleotides A and C. The position of A appeared in the sequence is $\{1,3,5\}$, and the position of C is $\{2,4,6\}$. We could calculate $\mu_A = 3$ and $\mu_C = 4$. Then the covariance between nucleotides A and C is

$$[(1-3)(2-4)/3 + (3-3)(4-4)/3 + (5-3)(6-4)/3]/10 = 4/15.$$

Secondly, we calculate the covariance between A and G. The position of G is $\{7,9\}$ and $\mu_G = 8$. The covariance between A and G is

$$\{[(1-2)(7-8)/2 + (3-2)(9-8)/2] + [(1-3)(7-8)/2 + (5-3)(9-8)/2]$$
$$+[(3-4)(7-8)/2 + (5-4)(9-8)/2]\}/(3 \times 10) = 2/15.$$

The covariances between the other pairs of nucleotides could be calculated in the same way.

Once we have obtained the covariances between pairs of nucleotides or amino acids, we integrate these covariances into the original natural vector of the sequence S. For nucleotides, there are $C_4^2 = 6$ possible pairs, and for amino acids, there are $C_{20}^2 = 190$ possible pairs. Consequently, the dimension of the extended natural vector for DNA sequences increases from 12 to 18, and for protein sequences, it expands from 60 to 250. This enhanced approach yields a novel type of natural vector that encapsulates additional natural statistical information for sequences. This advanced method effectively incorporates more comprehensive insights compared to the basic second-order natural vector, while it involves higher dimension, particularly in the case of protein sequences.

Chapter 7
Convex Hull Principle and Distinguishing Proteins from Arbitrary Amino Acid Sequences

7.1 The Convex Hull Principle

In every natural science field, it is important to discover the laws that govern it. In chemistry, there are three laws: the Law of Mass Conservation, the Law of Definite Proportions, and the Law of Multiple Proportions. In physics, there are Newton's Three Laws of Motion, which describe basic rules about how the motion of physical objects changes. Most of the scientific laws use mathematics to precisely describe natural phenomena. For instance, Maxwell formulated the four fundamental laws of electromagnetism (Gauss's law for electricity, Gauss's law for magnetism, Faraday's law, and Ampere's circuital law with Maxwell's correction) based on mathematical equations. Similarly, in molecular biology, an essential principle states that the biological properties of proteins or DNA are influenced by the distribution of the 20 amino acids or 4 nucleotides within their sequences. From a mathematical perspective, natural vectors corresponding to sequences from the same family tend to cluster in the high-dimensional space. To provide a more detailed explanation, we introduce the convex hull principle, which is founded on the mathematical concept of the convex hull.

The convex hull of a set A is the set containing all convex combinations of points in A. To be more precise, consider a set $A = \{a_1, a_2, \ldots, a_n\}$. The convex hull of A, denoted as $S(A)$, encompasses all points that can be expressed in the following form: $\sum_{i=1}^{n} \lambda_i a_i$, where $0 \leq \lambda_i \leq 1$ and $\sum_{i=1}^{n} \lambda_i = 1$. The convex hull principle means that the convex hull formed by natural vectors of biological sequences from one family does not intersect with the convex hull of natural vectors from another family. In this section, we will initially present methods to determine if two convex hulls intersect, followed by the validation of this principle using both protein and DNA sequences. Lastly, we will introduce a novel sequence detection framework based on the convex hull principle.

© The Author(s), under exclusive license to Springer Nature Switzerland AG 2023
S. S.-T. Yau et al., *Mathematical Principles in Bioinformatics*, Interdisciplinary
Applied Mathematics 58, https://doi.org/10.1007/978-3-031-48295-3_7

7.1.1 Methods for Determining Whether Two Convex Hulls Intersect

Consider two finite point sets $A = \{a_1, a_2, \ldots, a_n\}$ and $B = \{b_1, b_2, \ldots, b_m\}$ in $\mathbb{R}^k$, where S denotes the convex hull function. The convex hull intersection problem involves determining whether the two convex hulls $S(A)$ and $S(B)$ intersect. In this section, we present several mathematical approaches to solve the convex hull intersection problem. The detailed proofs of these methods can be found in [106].

7.1.1.1 The Projection-Line Method and the Normal Vector Method

An important property for convex hulls is that $S(A) \cap S(B) = \phi$ if and only if there is a line $l \subset \mathbb{R}^k$, s.t. $S(P_l(A)) \cap S(P_l(B)) = \phi$, where P_l is the projection operator to the line l.

The projection-line method states that if we can identify a line, denoted as P_l, such that the projections $P_l(A)$ and $P_l(B)$ of the point sets A and B onto this line are disjoint, then the convex hulls of the original point sets A and B do not intersect. This method significantly simplifies the computation by reducing the problem from a k-dimensional space to a one-dimensional space.

Since A and B are sets of finite points, the surfaces of $S(A)$ and $S(B)$ are composed of hyperplanes. In this case, we have a stronger property for convex hulls: The necessary and sufficient condition of $S(A) \cap S(B) = \phi$ is that there is a normal vector N of one hyperplane of $S(A)$ or $S(B)$, s.t. $S(P_N(A)) \cap S(P_N(B)) = \phi$, where P_N is the projection operator to the direction of the vector N.

Therefore, we can check all possible normal vectors to see whether two convex hulls intersect since the number of normal vectors for $S(A)$ and $S(B)$ is finite. It is called the normal vector method. The normal vector method can be seen as a special case of the projection-line method whose candidates of the projection lines are natural and finite.

7.1.1.2 The Subset Determination Method

In $\mathbb{R}^k$, the condition $S(A) \cap S(B) = \emptyset$ is necessary and sufficient when for all $i_1, i_2, \ldots, i_{k+1} \in [1, n]$ and $j_1, j_2, \ldots, j_{k+1} \in [1, m]$, we have $S(\{a_{i_1}, a_{i_2}, \ldots, a_{i_{k+1}}\}) \cap S(\{b_{j_1}, b_{j_2}, \ldots, b_{j_{k+1}}\}) = \emptyset$. Using this property, we can divide each convex hull into multiple convex blocks, each constructed by $k + 1$ points, and check for intersections among these smaller blocks. In a k-dimensional space, each convex block consists of $k + 1$ vertices and $k + 1$ faces formed by any possible combination of k vertices. The equations of each face and the corresponding normal vector of the convex block can be easily calculated. This enables us to employ the normal vector method to handle the intersection problem for subsets, resulting in a significant reduction in computation.

7.1.1.3 The Linear Programming Method

$S(A) \cap S(B) = \phi$ is equivalent with that there are no nonnegative real numbers λ_1, $\lambda_2, \ldots, \lambda_n, \mu_1, \mu_2, \ldots, \mu_m$ satisfying the following equations:

$$
\begin{cases}
\displaystyle\sum_{i=1}^{n} \lambda_i a_i = \sum_{j=1}^{m} \mu_j b_j. \\[2ex]
\displaystyle\sum_{i=1}^{n} \lambda_i = 1. \\[2ex]
\displaystyle\sum_{j=1}^{m} \mu_j = 1.
\end{cases}
\tag{7.1}
$$

By applying this theorem, we can convert the original problem into an algebraic problem. If any convex combination of the points in one set is equal to a convex combination of points in the other set, we can conclude that the two convex hulls intersect. Regardless of the dimension of the space or the number of points, this problem can be efficiently solved using linear programming functions available in various software. This method proves to be both time-saving and effective.

7.1.1.4 The Minimum Distance Method

Consider the optimization problem

$$
\begin{aligned}
\min \quad & \left\| \sum_{i=1}^{n} \lambda_i a_i - \sum_{j=1}^{m} \mu_j b_j \right\| \\
s.t. \quad & \sum_{i=1}^{n} \lambda_i = 1. \\
& \sum_{j=1}^{m} \mu_j = 1. \\
& \lambda_i \geq 0, \ i = 1, \ldots, n, \\
& \mu_j \geq 0, \ j = 1, \ldots, m.
\end{aligned}
\tag{7.2}
$$

The condition $S(A) \cap S(B) = \emptyset$ is equivalent to the minimum value of an optimization problem being greater than zero. In this case, we can reframe the problem as calculating the minimum distance between the two convex hulls. If the minimum distance is positive, it indicates that the convex hulls are disjoint. Various mathematical software packages provide quadratic programming functions that can efficiently solve this minimization problem.

7.1.2   The Verification of the Convex Hull Principle

In this part, we will use biological data to verify the convex hull principle. To be more specific, we will show that biological sequences from different families can be transformed into pairwise disjoint convex hulls by the natural vector method.

7.1.2.1   The Verification by Protein Sequences

We first show that the convex hull principle holds for protein sequences [107]. Two protein datasets are considered. The first dataset is the protein kinase dataset including 31,355 sequences, which can be divided into 107 families. The second dataset is the human protein dataset including 10,983 sequences that are divided into 2156 families. We use the natural vector method with covariance introduced in Sect. 6.6 to construct convex hulls. The natural vector method with covariance transforms protein sequences into 250-dimensional vectors. By applying the linear programming method as mentioned earlier, it is observed that there is no intersection between any pair of convex hulls in both datasets. This observation further confirms the validity of the convex hull principle. (There are 5671 pairs of convex hulls in the first dataset and 2,323,090 pairs of convex hulls in the second dataset.)

To visualize the results, we employ the linear discriminant analysis (LDA) method for dimension reduction [109]. LDA is utilized to determine the linear separability of two groups, reducing the vector dimension from 250 to 2. In Figs. 7.1 and 7.2, we present visualizations of a pair of convex hulls from two datasets, respectively. Notably, the protein space points exhibit clustering rather than a dispersed distribution. This observation implies that as new protein kinase sequences are incorporated, their points are expected to lie within the approximate convex hull of known protein kinase families. Furthermore, Fig. 7.3 showcases fifteen families within the AGC group of human protein kinases, aiding in our comprehension of the disjoint nature of distinct convex hulls.

7.1.2.2   The Verification by DNA Sequences

The convex hull principle also holds for DNA sequences [108]. The dataset for verification is a viral genome dataset including 7382 sequences from 83 families. Different from what we have done when dealing with protein sequences, we use the standard natural vector method without covariance but with high orders to construct convex hulls. The reason is that the method in Sect. 6.6 transforms DNA sequences into vectors of only 18 dimensions, which may not include sufficient information.

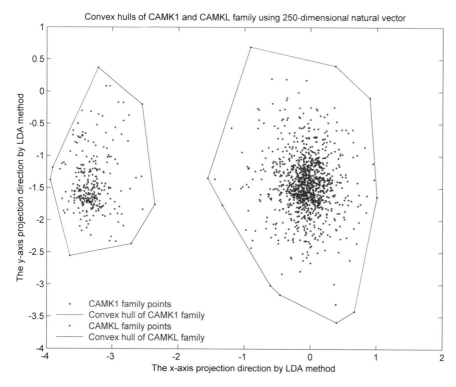

Fig. 7.1 Convex hulls of the CAMK1 (including 295 sequences) and CAMKL (including 1075 points) animal protein kinase families after dimension reduction by LDA method

Therefore, we consider the standard natural vectors with high-order moments to separate different families. The specific order depends on the dataset. For the viral genome dataset considered in [108], the order for the complete separation is seven. In other words, 3403 pairs of convex hulls are disjoint after transforming the DNA sequences into order 7 natural vectors in a 32-dimensional space.

7.1.3   New Sequence Detection

Despite the rapid growth in the number of known genomes and protein sequences, they represent only a small fraction of the vast diversity found in nature. Biologically, the detection and prediction of new genome or protein sequences based on real sequence data remains a challenging and crucial problem. The convex hull principle provides us with a new framework to solve this problem. Since the natural vector of a new sequence must lie in the convex hull of its family, the problem of detecting new sequences can be transformed into finding sequences whose corresponding natural vector is in a given convex hull. In this section, we will first introduce the method to

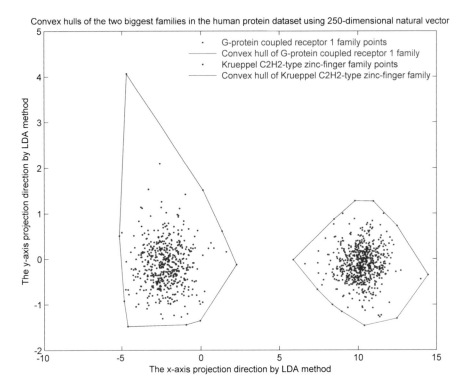

Fig. 7.2 Convex hulls of the G-protein coupled receptor 1 (including 670 sequences) and Krueppel C2H2-type zinc-finger (including 538 sequences) after dimension reduction by LDA method

determine the composition of biological sequences and then introduce two heuristic methods to detect new biological sequences. Without loss of generality, we focus on genome sequences, and all contents can be transferred to protein sequences without difficulty.

7.1.3.1 Determination of the Nucleotide Composition of Genome Sequences

In this part, we will discuss the constraints on the first four elements of natural vectors of genomes, which describe the nucleotide composition of genome sequences. An obvious observation is that the first four elements of a natural vector should be integers. In fact, there is a tighter constraint shown in [110].

For an order 2 natural vector, it is not hard to prove that its elements satisfy the following equations:

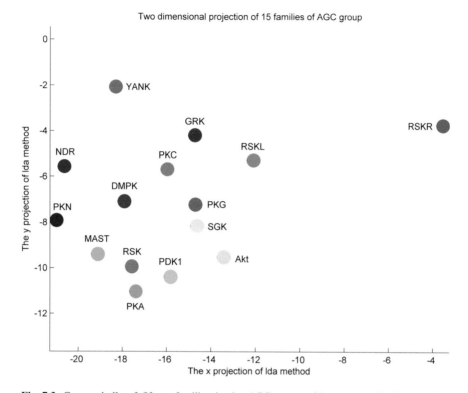

Fig. 7.3 Convex hulls of fifteen families in the AGC group of human protein kinases after dimension reduction by LDA method

$$\sum_{k\in K} n_k \mu_k = \sum_{j=1}^{n} j,$$

$$\sum_{k\in K} \sum_{i=1}^{n_K} s[k][i]^2 = \sum_{k\in K} nn_k D_2^k + \sum_{k\in K} n_k \mu_k^2 = \sum_{j=1}^{n} j^2, \tag{7.3}$$

where $K = \{A, T, C, G\}$ for DNA sequences.

Therefore, for a real natural vector

$$(n_{A,\alpha}, n_{C,\alpha}, n_{G,\alpha}, n_{T,\alpha}, \mu_{A,\alpha}, \mu_{C,\alpha}, \mu_{G,\alpha}, \mu_{T,\alpha}, D_2^{A,\alpha}, D_2^{C,\alpha}, D_2^{G,\alpha}, D_2^{T,\alpha}),$$

which lies in the convex hull of N natural vectors

$$(n_{A,i}, n_{C,i}, n_{G,i}, n_{T,i}, \mu_{A,i}, \mu_{C,i}, \mu_{G,i}, \mu_{T,i}, D_2^{A,i}, D_2^{C,i}, D_2^{G,i}, D_2^{T,i}),$$

where $i = 1, \ldots, N$, there exists a solution about α_i in the following equations:

$$
\begin{cases}
\sum_{i=1}^{N} \alpha_i = 1, \alpha_i \geq 0 \\
n_{k,\alpha} = \sum_{i=1}^{N} \alpha_i n_{k,i} \\
\sum_{k \in K} n_K \left(\sum_{i=1}^{N} \alpha_i \mu_{k,i} \right) = \frac{1}{2} n_\alpha (n_\alpha + 1) \\
\sum_{k \in K} n n_k \sum_{i=1}^{N} \alpha_i D_2^{k,i} + n_k \left(\sum_{i=1}^{N} \alpha_i \mu_{k,i} \right)^2 = \frac{1}{6} n_\alpha (n_\alpha + 1)(2n_\alpha + 1) \\
n_\alpha = \sum_{k \in K} n_{k,\alpha}
\end{cases}
\tag{7.4}
$$

Equation (7.4) is hard to solve directly because this is a large-scale underdetermined equation of α_i. Therefore, in real applications, we first fix a four-dimensional integer point $(n_{k,\alpha})_{k \in K}$ and define $n_\alpha = \sum_{k \in K} n_{k,\alpha}$. Then we transform the existence of the solution of Eq. (7.4) to an optimization problem.

$$
\begin{cases}
\min_{\vec{\alpha}} (\text{or } \max_{\vec{\alpha}}) \quad \sum_{k \in K} n_\alpha n_{k,\alpha} \sum_{i=1}^{N} \alpha_i D_2^{k,i} + n_{k,\alpha} \left(\sum_{i=1}^{N} \alpha_i \mu_{k,i} \right)^2 \\
s.t. \quad \sum_{i=1}^{N} \alpha_i = 1 \\
\quad \sum_{i=1}^{N} \alpha_i n_{k,i} = n_{k,\alpha} \quad \forall k \in K \\
\quad \sum_{k \in K} n_{k,\alpha} \left(\sum_{i=1}^{N} \alpha_i \mu_{k,i} \right) = \frac{n_\alpha(n_\alpha+1)}{2} \\
\quad 0 \leq \alpha_i \leq 1, \quad i = 1, 2, \cdots, N
\end{cases}
\tag{7.5}
$$

A necessary condition of a four-dimensional integer point to be the first four elements of a natural vector is that the target value $\frac{1}{6} n_\alpha (n_\alpha + 1)(2n_\alpha + 1)$ lies between the minimum and the maximum of the corresponding optimal problem. According to this condition, we can choose the candidates of the first four elements of a natural vector in two steps. First, we find the integer points of a given convex hull. Second, we check the necessary condition to determine the final candidates. There are proven calculation methods for both two steps [110].

7.1.3.2 Heuristic Methods to Detect New Sequences

Solving the new genome detection problem as an optimization problem is proven to be NP-hard [111], making it challenging to find a complete solution. Therefore, heuristic methods are commonly employed. In a previous discussion, we presented a method to estimate the nucleotide counts in a genome sequence. The objective of heuristic methods is to generate a sequence that has a corresponding natural vector within the convex hull of its family, given the nucleotide counts. In this section, we will introduce two approaches: the Random-permutation Algorithm with Penalty (RAP) and the Random-permutation Algorithm with Penalty and Constrained Search (RAPCOS) [111].

The idea of RAP is natural. After determining the counts of nucleotides, we can generate a sequence randomly according to the given counts. Then we randomly conduct a permutation, and the new sequence is recorded only when a loss function that describes how far the present status is from the goal decreases. If we know a real target natural vector that lies in the given convex hull and we want to find a corresponding sequence for it, we can take the loss function as $loss(S^{seq}) = d(v^{seq}, v^{tg})$, where v^{seq} is the input of the function and v^{tg} is the target natural vector. We can also let v^{tg} be a fuzzy natural vector, which may not actually be a natural vector but have some properties of natural vectors. For example, a 12-dimensional vector whose corresponding Eq. (7.4) has solutions can be regarded as a fuzzy natural vector.

In algebra, it is widely acknowledged that any permutation can be expressed as a combination of 2 cycles permutations, i.e., transpositions. In simpler terms, we can convert one sequence into another, provided both sequences have the same count of nucleotides, using a series of transpositions. Hence, it is reasonable for us to focus exclusively on 2 cycle permutations. During each step of a random 2 cycle permutation, the following actions are taken:

1. Randomly select two nucleotides, denoted as k and q, from the set of nucleotides $K = \{A, T, C, G\}$, where $k \neq q$.
2. Choose two positions from the positions occupied by nucleotides k and q, respectively. Subsequently, perform the 2 cycle permutation by exchanging the nucleotides at the selected positions if it leads to a reduction in the loss function.

By iteratively following these steps, the random 2 cycle permutation enables exploration of various nucleotide arrangements, with the objective of minimizing the associated loss function.

To improve the efficiency of selecting two nucleotides, a penalty probability is utilized, which assigns varying probabilities to different nucleotides. Specifically, for a given nucleotide, the likelihood of selecting it for permutation increases as the difference between its current natural vector and the target natural vector grows larger. In order to account for fixed counts associated with a particular nucleotide, both the mean position and the second-order central normalized moment are measured simultaneously. This implies that we consider the statistical properties of the nucleotide's distribution in addition to its frequency of occurrence.

$$P_k(v^{seq}, v^{tg}) = \frac{(|\mu_k^{seq} - \mu_k^{tg}| + |D_2^{k,seq} - D_2^{k,tg}|)}{\sum_{j \in K} (|\mu_j^{seq} - \mu_j^{tg}| + |D_2^{j,seq} - D_2^{j,tg}|)}. \tag{7.6}$$

The algorithm flowchart of RAP is shown in Algorithm 1.

Algorithm 1 Random-permutation Algorithm with Penalty (RAP)

Require: the target natural vector $\mathbf{v}^{tg}$, the maximal iteration number M, and a preset limit ε.
1: Get the nucleotide composition based on $\mathbf{v}^{tg}$ and randomly generate a sequence $\mathbf{S}^{seq}$. Let Iter $= 0$, $\mathbf{S}^{new} = \mathbf{S}^{seq}$.
2: **while** $\text{loss}(\mathbf{S}^{seq}) > \varepsilon$ and iter $< M$ **do**
3:    Iter = Iter + 1.
4:    **while** $\text{loss}(\mathbf{S}^{new}) \geq \text{loss}(\mathbf{S}^{seq})$ **do**
5:        Randomly Select $k, q \in K$ ($k \neq q$) based on the probability in Eq. (7.6).
6:        Randomly Select pos_k and pos_q from the positions of k and q, respectively.
7:        Get $\mathbf{S}^{new}$ from $\mathbf{S}^{seq}$ (Do a permutation between pos_k and pos_q).
8:    **end while**
9:    Let $\mathbf{S}^{seq} = \mathbf{S}^{new}$.
10: **end while**
11: The $\mathbf{S}^{seq}$ is the output of the algorithm.

In RAP's permutation, it often requires multiple iterations to discover a new sequence that minimizes the loss function. However, we can enhance the algorithm's efficiency by imposing constraints on the selection of nucleotides and their positions. During each step of the 2 cycle permutation, we choose two distinct nucleotides, denoted as k and q, from a set K. Following the permutation, the values of μ_k, μ_q, D_2^k, and D_2^q undergo changes. Our objective is to ensure that these four values simultaneously approach the target natural vector more closely than in the previous iteration. In the subsequent discussion, we will delve into the analysis of the alterations in the natural vector subsequent to the permutation.

It is not hard to check the following result:

Lemma 7.1 *If for $k \in K$, $\mu_k^{seq} < \mu_k^{tg}$, there must exist $q \in K, q \neq k$, where $\mu_q^{seq} > \mu_q^{tg}$ and vice versa.*

By each 2 cycle permutation, if μ_k increases, then μ_q decreases and vice versa. So it is a good idea to choose two nucleotides whose mean positions have the opposite relation with those of the target natural vector. According to this strategy, we can classify the choices of two nucleotides into eight cases. We use the letter G to represent "Greater than that of the target natural vector" and use the letter L to represent "Lower than that of the target natural vector." For example, the case LL&GL means that $\mu_k^{seq} < \mu_k^{tg}$, $D_2^{k,seq} < D_2^{k,tg}$, $\mu_q^{seq} > \mu_q^{tg}$, $D_2^{q,seq} < D_2^{q,tg}$.

Therefore, we have eight cases denoted as LL&GL, LL&GG, LG&GL, LG&GG, GL&LL, GL&LG, GG&LL, GG&LG. We can set a constraint for each case to ensure the reduction of the loss function. (In the statement of Theorem 7.1, we sometimes use μ_k to represent μ_k^{seq} to make the formulas more compact.)

Theorem 7.1 *For a genome sequence $\mathbf{S}^{seq}$ and a target sequence $\mathbf{S}^{tg}$, the corresponding natural vectors are $\mathbf{v}^{seq}$ and $\mathbf{v}^{tg}$, respectively. Denote the chosen position in a permutation as s_k, s_q. Denote $h_k(x) = (n_k - 1)^{-1}[2n_k\mu_k - (n_k + 1)x]$, $g_k(x) = (n_k + 1)^{-1}[2n_k\mu_k - (n_k - 1)x]$, where $k \in K$ can be replaced by $q \in K$. And denote $\rho_1 = (n_k + n_q)^{-1}[(n_q + 1)n_k\mu_k - (n_k - 1)n_q\mu_q]$. Then the following* *requirement ensures that* $\|\mathbf{v}^{new} - \mathbf{v}^{tg}\| < \|\mathbf{v}^{seq} - \mathbf{v}^{tg}\|$ *for each case. Furthermore, it is guaranteed that* $|\mu_k^{new} - \mu_k^{tg}| < |\mu_k^{seq} - \mu_k^{tg}|$, $|\mu_q^{new} - \mu_q^{tg}| < |\mu_q^{seq} - \mu_q^{tg}|$, $|D_2^{k,new} - D_2^{k,tg}| < |D_2^{k,seq} - D_2^{k,tg}|$, $|D_2^{q,new} - D_2^{q,tg}| < |D_2^{q,seq} - D_2^{q,tg}|$.

LL&GL: $\mu_k < \mu_q$, $\rho_1 < s_k < \mu_q$, $\max\{s_k, h_k(s_k)\} < s_q < g_q(s_k)$.

LL&GG: $s_q > \max\{s_k, h_k(s_k), g_q(s_k), \mu_q\}$.

LG&GL: $s_k < \min\{\mu_k, \mu_q\}$, $s_k < s_q < \min\{h_k(s_k), g_q(s_k)\}$.

LG&GG: $s_k < \min\{\mu_k, \rho_1, g_k(\mu_q)\}$, $\max\{s_k, \mu_q, g_q(s_k)\} < s_q < h_k(s_k)$.

GL&LL: $\mu_k > \mu_q$, $\mu_q < s_k < \rho_1$, $g_q(s_k) < s_q < \min\{s_k, h_k(s_k)\}$.

GL&LG: $s_q < \min\{s_k, h_k(s_k), g_q(s_k), \mu_q\}$.

GG&LL: $s_k > \max\{\mu_k, \mu_q\}$, $\max\{h_k(s_k), g_q(s_k)\} < s_q < s_k$.

GG&LG: $s_k > \max\{\mu_k, \rho_1, g_k(\mu_q)\}$, $h_k(s_k) < s_q < \min\{s_k, \mu_q, g_q(s_k)\}$.

According to Theorem 7.1, RAP method can be accelerated by the constrained search. The framework of the algorithm is the same, while there are constraints when selecting the nucleotides and the positions. When selecting nucleotides k, q, we need $(\mu_k^{seq} - \mu_k^{tg})(\mu_q^{seq} - \mu_q^{tg}) > 0$. When selecting the positions, constraints in Theorem 7.1 are considered. By these constraints, we can accelerate the previous RAP method. It is important to note that the constraints outlined in Theorem 7.1 may be overly stringent in certain cases. Specifically, when the loss function is extremely close to its minimum value, it is possible that no positions satisfy the given requirements. In such situations, we revert to the original strategy of the RAP method, which involves random permutation, if the constrained search fails to yield any results.

The experiments conducted in [111] demonstrate the practicality of both methods, RAP and RAPCOS, with RAPCOS exhibiting superior performance compared to RAP when the loss function is defined as the distance between the current natural vector and the target natural vector.

7.2 Distinguishing Proteins from Arbitrary Amino Acid Sequences

Protein sequences are composed of amino acids and can be identified as valid protein sequences if they meet certain criteria. While known proteins represent only a small fraction of all possible combinations of amino acids, efforts have been made to understand the structure of existing protein sequences in order to define the protein universe. However, we still need some criterion for determining whether an arbitrary amino acid sequence can be classified as a protein.

It was discovered that when arbitrary amino acid sequences are examined within a suitable geometric framework, protein sequences tend to cluster together. This observation has led to the development of a new computational test. Remarkably, this test has demonstrated a high level of accuracy in determining whether an arbitrary amino acid sequence has the characteristics of a protein.

This section presents a computational test, as described in [112], with an accuracy of 99.69% for determining the protein potential of an arbitrary amino acid sequence. Our test provides a rapid means to assess whether a given amino acid sequence has the characteristics of a protein. It relies on the observation that natural vectors of genuine protein sequences exhibit clustering behavior, aligning with the principle of convex hull.

7.2.1   The Principle and the Algorithm

For each amino acid sequence, we consider the 60-dimensional natural vector to represent this property:

$$(n_A, n_R, \ldots, n_V, \mu_A, \mu_R, \ldots, \mu_V, D_2^A, D_2^R, \ldots, D_2^V). \tag{7.7}$$

To begin, we establish the one-to-one nature of the mapping. While theoretically, we cannot guarantee one-to-one correspondence due to insufficient order, this is not the case with actual protein sequences. The primary outcome of this section is the pivotal finding that, by gathering all known and reviewed complete protein sequences submitted prior to March 6, 2013, from the UniprotKB database [113], and calculating their natural vectors, we confirm that the 60-dimensional natural vector representation indeed exhibits one-to-one correspondence with known protein sequences.

To gain insight into the distribution of protein space within the amino acid sequence space, we employ a visualization technique by plotting the points along two coordinate axes associated with the amino acid Alanine (A). Figure 7.4a and b depicts the two-dimensional projection onto the (n_A, μ_A) coordinate plane, while c and d illustrates its projection onto the (n_A, D_2^A) coordinate plane. These visual representations provide a clearer understanding of how the points in protein space are distributed relative to the amino acid sequence space.

In the (n_A, μ_A) coordinate plane, the projection of amino acid space is bounded by the following three lines:

$$\mu_{A,min} : \mu_A = \frac{n_A + 1}{2} \tag{7.8}$$

$$\mu_{A,max} : \mu_A = \frac{2n + 1 - n_A}{2} \tag{7.9}$$

$$n_A = 0. \tag{7.10}$$

In the (n_A, D_2^A) coordinate plane, the projection of the amino acid space is bounded by the following three curves:

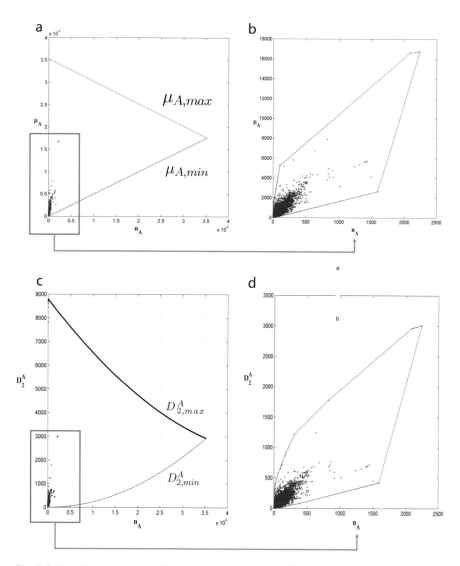

Fig. 7.4 The Alanine convex hull was computed using the Uniprot 2013_03 dataset. The blue points in each of the four subfigures represent vectors corresponding to proteins. Subfigure (**a**) illustrates the projection onto the (n_A, μ_A) coordinate plane, while subfigure (**c**) depicts the projection onto the (n_A, D_2^A) coordinate plane. Subfigure (**b**) provides an enlarged view of the protein area in (**a**), with black lines indicating the boundaries of the convex hull. Similarly, subfigure (**d**) presents an enlarged view of the protein area in (**c**), with black lines representing the boundaries of the convex hull

$$D_{2,min}^A : D_2^A = \frac{n_A^2 - 1}{12n} \tag{7.11}$$

$$
D^A_{2,max} : D^A_2 = \begin{cases} \dfrac{n - n_A}{4} + \dfrac{n^2_A}{12n} - \dfrac{1}{12n} & \text{if } n_A \text{ is even} \\[3mm] \dfrac{(n_A - 1)(n_A + 1)(n - n_A)}{4n^2_A} + \dfrac{n^2_A}{12n} - \dfrac{1}{12n} & \text{if } n_A \text{ is odd} \end{cases}
$$

$$
\tag{7.12}
$$

$$
n_A = 0. \tag{7.13}
$$

We will prove these boundaries in the final part of this chapter.

In these equations, n represents the maximum length of proteins in the dataset. For the collection of known, reviewed complete protein sequences found in the UniprotKB data as of March 6, 2013, it is determined that $n = 35{,}213$.

From Fig. 7.4, it is evident that the points in the protein space are clustered rather than widely distributed. This observation leads us to believe that as new protein sequences are discovered, their corresponding points will likely fall within the approximate boundaries of the convex hull formed by the points corresponding to known protein sequences.

Although we can directly deal with the convex hull in the 60-dimensional space, it is more efficient to deal with the convex hull of (n_k, μ_k, D^k_2) for each amino acid k in the 3-dimensional space. We use the word the k-protein area to represent the set of points (n_k, μ_k, D^k_2).

Several tests have been conducted to demonstrate the stability of the boundaries of these protein areas over time, even as more proteins are described. These tests serve as robust evidence supporting the validity of the following computational test.

In order to determine if an arbitrary amino acid sequence can be classified as a protein sequence, the following steps are undertaken:

1. The natural vector of the amino acid sequence is computed.
2. A pre-computed vector database containing natural vectors corresponding to known protein sequences is established.
3. If the natural vector of the amino acid sequence is found within the vector database, we check whether the amino acid sequence is a known sequence. It is concluded that the amino acid sequence is a protein only if it is identical to a known protein sequence, as a result of the one-to-one correspondence between natural vectors and protein sequences.
4. In the event that the natural vector of the amino acid sequence is not in the vector database, we need to verify the inclusion of each of the 20 points (n_k, μ_k, D^k_2) in a 3-dimensional space within their respective convex hulls for the corresponding k-protein areas. If all of these checks prove successful, it can be concluded that this amino acid sequence has the potential to be categorized as a real protein. Conversely, if not all points lie within their corresponding convex hulls, the second key result suggests that the amino acid sequence is unlikely to be a protein sequence.

7.2.2 The Verification by Real Protein Sequences

To validate our algorithm, we utilize three snapshots of the UniprotKB database: Uniprot 2013_03 (March 6, 2013), Uniprot 2014_03 (March 19, 2014), and Uniprot 2014_06 (June 11, 2014). In each instance, we exclusively consider the reviewed and complete proteins for analysis and evaluation.

We employ a consistent methodology to generate all three datasets. We include the keyword "Complete proteome[KW-0181]" to exclusively select complete sequences, thereby eliminating any sequences with missing amino acids. Additionally, we incorporate the keyword "Reviewed" to narrow down the dataset to only reviewed proteins. Redundant sequences are not reduced during this process.

Following the download, the datasets undergo normalization by removing protein sequences that contain Selenocysteine (U) and Pyrrolysine (O), as well as sequences containing placeholders (B, Z, J, X). The resulting datasets are then subjected to a cleaning process. Table 7.1 presents the sequence numbers for each dataset after this cleaning process:

To verify the one-to-one correspondence for 60-dimensional natural vectors, we utilize the Uniprot 2013_03 and Uniprot 2014_03 datasets. The analysis reveals that the number of unique protein sequences in each dataset is equivalent to the number of distinct natural vectors. This finding suggests that there is a direct mapping between protein sequences and the corresponding 60-dimensional natural vectors in both datasets.

We conducted multiple tests to assess the effectiveness of our computational method in determining the proteinogenic nature of a given amino acid sequence.

Initially, we utilized the 391,704 distinct protein sequences present in both Uniprot 2013_03 and Uniprot 2014_03 datasets to generate twenty 3-dimensional convex hulls using the Qhull algorithm in MATLAB software. Subsequently, we examined the 3810 sequences exclusively present in Uniprot 2014_03 to determine how many of them fell outside the boundaries of the twenty convex hulls. Surprisingly, only 14 sequences (equivalent to a mere 0.37% of the 3810

Table 7.1 Detailed counts of sequences in the three snapshots of UniprotKB

Number of distinct sequences in Uniprot 2013_03	392,455
Number of distinct sequences in Uniprot 2014_04	395,514
Number of distinct sequences in Uniprot 2014_06	397,348
Number of distinct sequences in Uniprot 2013_03 and Uniprot 2014_04	391,704
Number of distinct sequences contained in all three datasets	391,528
Number of sequences in Uniprot 2013_03 but not Uniprot 2014_03	751
Number of sequences in Uniprot 2014_03 but not Uniprot 2013_03	3810
Number of sequences in Uniprot 2013_03 and Uniprot 2014_03, but not Uniprot 2014_06	176
Number of sequences in Uniprot 2014_06 but not in the intersection of Uniprot 2013_03 and Uniprot 2014_03	5820

Table 7.2 The 14 protein sequence outliers in Uniprot 2014_03 and their distances from the convex hulls

No.	Sequence length	Access ID	Convex hull the sequences fall outside	Distance to convex hull
1	11	P85817	Asparagine (N)	0.0177
2	16	P81071	Aspartic acid (D)	0.0110
3	19	P68116	Aspartic acid (D)	0.0018
4	20	P14469	Isoleucine (I)	0.003
5	199	Q9ZVZ9	Histidine (H)	0.0000208
6	211	P33191	Tyrosine (Y)	0.0027
7	237	Q6M923	Glutamine (Q)	0.0362
8	287	P50751	Proline (P)	0.0044
9	392	Q5A8I8	Proline (P)	33.9023
10	1086	Q59XL0	Methionine (M)	0.4508
11	1129	Q9QR71	Glutamic acid (E)	5.5955
			Glutamine (Q)	1.4455
12	1404	Q59SG9	Serine (S)	0.2427
13	2346	A1Z8P9	Glycine (G)	0.2179
14	3461	P62288	Arginine (R)	1.8593

sequences) were found to lie outside one of the convex hulls. It is worth noting that none of these 14 sequences was significantly distant from the convex hull boundaries. The details of these sequences, along with their respective distances from the convex hulls, are presented in Table 7.2. Furthermore, Fig. 7.5 visually illustrates one of the 14 sequences positioned outside a convex hull. Additionally, we repeated the process using all 392,455 sequences from Uniprot 2013_03, and the results remained consistent.

Next, we proceeded to calculate new convex hulls using the aforementioned 391,704 protein sequences combined with the 14 sequences that did not pass the first test. Subsequently, we examined each of the 5820 sequences present in Uniprot 2014_06 but not found in the intersection of Uniprot 2013_03 and Uniprot 2014_03 to determine how many of them failed to fall within any of the convex hulls. Remarkably, only 18 sequences (equivalent to a mere 0.31% of the 5820 sequences) were found to lie outside one of the convex hulls. Similar to the previous test, none of these 18 sequences were significantly distant from the convex hull boundaries. Table 7.3 provides a detailed list of these 18 sequences along with their respective distances from their corresponding convex hulls.

In the final test, we examined whether the protein sequences from Top 7 [114], HOP2 [115], and GLUT1 [116] were located within the convex hulls constructed using the sequences present in both Uniprot 2013_03 and Uniprot 2014_03. As anticipated, all three proteins were found to lie inside the twenty convex hulls. This result aligns with our expectations.

Indeed, the results of our tests are promising. Although there were a few proteins that fell outside at least one of the convex hulls, the percentage of proteins

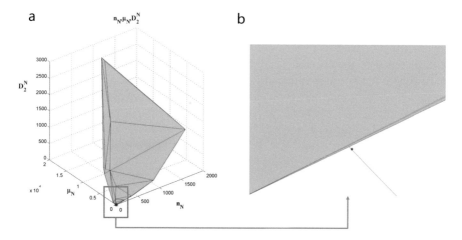

Fig. 7.5 The protein sequence with Access ID P85817 is identified as lying outside the convex hull corresponding to Asparagine. In subfigure (**a**), the cyan surfaces represent the convex hulls in 3-dimensional space, while the red point represents the coordinate n_N, μ_N, D_2^N for this particular sequence. Subfigure (**b**) provides an enlarged view of subfigure (**a**), clearly demonstrating that the point falls outside the boundaries of the convex hull

Table 7.3 The 18 protein sequence outliers from Uniprot 2014_06 and their distances from the convex hulls

No.	Sequence length	Access ID	Convex hull the sequences fall outside	Distance to convex hull
1	20	P82867	Aspartic acid (D)	0.0045
2	19	P68214	Aspartic acid (D)	0.0309
3	15	P80612	Alanine (A)	0.3850
4	267	P14918	Arginine (R)	0.0000042949
5	150	P27787	Phenylalanine (F)	0.00025511
6	372	Q5AKU5	Histidine (H)	0.00013742
7	105	Q2RB28	Leucine (L)	0.000053612
8	105	B9GBM3	Leucine (L)	0.000053612
9	94	Q5G8Z3	Aspartic acid (D)	0.0022
10	391	P46525	Serine (S)	0.1044
11	838	P08489	Methionine (M)	0.0840
			Valine (V)	3.0627
12	848	P10388	Glutamic acid (E)	0.1269
			Methionine (M)	0.1408
			Valine (V)	3.1465
13	240	P04702	Glycine (G)	0.000036147
14	240	P06677	Glycine (G)	0.000036147
15	240	P04703	Glycine (G)	0.000036147
16	240	P06676	Glycine (G)	0.000036147
17	267	P04698	Glycine (G)	0.000070559
18	187	B6U769	Proline (P)	0.0037

failing the test is relatively small in both cases (0.37% and 0.31%). Furthermore, it is noteworthy that the proteins not contained within a convex hull are never significantly distant from the boundary of the convex hull. This suggests that as more protein sequence data become available, the computed convex hulls will become increasingly reliable and eventually stabilize, providing an accurate and efficient method for determining whether a given amino acid sequence can be classified as a protein sequence. These findings provide a strong indication that the approach holds great potential for practical applications in protein sequence analysis.

We anticipate that researchers in two distinct domains will employ this test. Protein designers seeking to synthesize novel proteins can utilize this test to rapidly identify amino acid sequences that are unlikely to fold into functional proteins, thus saving resources on costly laboratory synthesis efforts. Likewise, biologists investigating alternative splicing [117] now possess a novel tool for predicting the production of authentic protein sequences resulting from alternative splicing events.

7.2.3 Derivation for the Equations of the Boundaries of Amino Acid Space

Finally, we derive the equations of the boundaries introduced before rigorously.

7.2.3.1 The Boundaries of Amino Acid k in the (n_k, μ_k) Plane

Our objective is to determine the minimum value $\mu_{k,\min}$ and maximum value $\mu_{k,\max}$ of the mean for sequences of length n that contain n_k instances of the amino acid k.

Theorem 7.2 *Let n_k be the number of occurrences of amino acid k in a sequence of length n. Then:*

(a)

$$\mu_{k,min} = \frac{n_k + 1}{2}. \tag{7.14}$$

(b)

$$\mu_{k,max} = \frac{2n - n_k + 1}{2}. \tag{7.15}$$

Proof

(a) Obviously, if we choose the amino acid k to be distributed in positions $x_1 = 1$, $x_2 = 2, \ldots, x_{n_k} = n_k$, then we will get the minimum value of μ_k that is

$$\mu_{k,min} = \frac{1 + 2 + \cdots + n_k}{n_k} = \frac{n_k + 1}{2}. \tag{7.16}$$

(b) Similarly, if we choose the amino acid k to be distributed in positions $x_1 = n + 1 - n_k$, $x_2 = n + 2 - n_k, \ldots, x_{n_k} = n$, then we will get the maximum value of μ_k that is

$$\begin{aligned}
\mu_{k,max} &= \sum_{i=1}^{n_k} \frac{n + i - n_k}{n_k} \\
&= \frac{n_k(n - n_k) + \frac{n_k(n_k + 1)}{2}}{n_k} \\
&= \frac{2n - n_k + 1}{2}.
\end{aligned} \tag{7.17}$$

Remark 7.1   The equation $\mu_k = \frac{n_k + 1}{2}$ represents the lower boundary of the region, where $1 \le n_k \le c$. Similarly, the equation $\mu_k = \frac{2c - n_k + 1}{2}$ represents the upper boundary of the region, where $1 \le n_k \le c$. In these equations, c represents the maximum length of the sequences in our dataset.

7.2.3.2   The Boundaries of Amino Acid k in the (n_k, D_2^k) Plane

To begin, we aim to find the maximum value $D_{2,max}^k$ of the second normalized moment for sequences of length n that contain n_k instances of the amino acid k.

Lemma 7.2   *Let $\mu(\cdot)$ be the average function. Define*

$$F(x) = (x - \mu(x, x_1, \ldots, x_m))^2 + \sum_{i=1}^{m}(x_i - \mu(x, x_1, \ldots, x_m))^2.$$

Then $F(x) > F(y)$ if and only if

$$|x - \mu(x_1, \ldots, x_m)| > |y - \mu(x_1, \ldots, x_m)|.$$

(x is not necessary to be an integer in this lemma.)

Proof

$$\begin{aligned}
F'(x) &= \frac{2m}{m+1}(x - \mu(x, x_1, \ldots, x_m)) - \frac{2}{m+1}\sum_{i=1}^{m}(x_i - \mu(x, x_1, \ldots, x_m)) \\
&= 2(x - \mu(x, x_1, \ldots, x_m))
\end{aligned}$$

$$\tag{7.18}$$

$$F''(x) = 2\frac{m}{m+1} > 0. \tag{7.19}$$

Since $F'(x) = 0 \iff x = \mu(x, x_1, \ldots, x_m) \iff x = \mu(x_1, \ldots, x_m)$, we know that $F(x)$ increases when $x > \mu(x_1, \ldots, x_m)$ and decreases when $x < \mu(x_1, \ldots, x_m)$. Furthermore, from the form of $F'(x)$, it is not hard to check that $F(x)$ is symmetric about the line $x = \mu(x_1, \ldots, x_m)$. Therefore, $F(x) > F(y)$ if and only if $|x - \mu(x_1, \ldots, x_m)| > |y - \mu(x_1, \ldots, x_m)|$.

Now, we can determine the specific distribution of the amino acid k where the maximum value of the second normalized moment is assumed.

Theorem 7.3 *Let $0 < n_k \le n$ be fixed positive integers. Consider a distribution of the amino acid k given by $1 \le x_1 < x_2 < \cdots < x_{n_k} \le n$. If the second normalized moment $D_2^k(x)$ achieves the maximum value $D_{2,max}^k$ among all possible distributions, then the values $x_1, x_2, \ldots, x_{n_k}$ follow a specific pattern.*

Case 1: n_k even integer:

$$x_1 = 1, x_2 = 2, \ldots, x_{\frac{n_k}{2}} = \frac{n_k}{2}, x_{\frac{n_k}{2}+1} = n + 1 - \frac{n_k}{2},$$
$$x_{\frac{n_k}{2}+2} = n + 2 - \frac{n_k}{2}, \ldots, x_{n_k} = n. \tag{7.20}$$

Case 2: n_k odd integer:

$$x_1 = 1, x_2 = 2, \ldots, x_{\frac{n_k-1}{2}} = \frac{n_k-1}{2}, x_{\frac{n_k+1}{2}+1} = \frac{n_k+1}{2},$$
$$x_{\frac{n_k+1}{2}+1} = n + 1 - \frac{n_k-1}{2}, x_{\frac{n_k+1}{2}+2} = n + 2 - \frac{n_k-1}{2}, \ldots, x_{n_k} = n. \tag{7.21}$$

Proof We will focus on proving the case when n_k is an even integer, as the proof for the case when n_k is an odd integer follows a similar logic.

We first show that the amino acid k piles on the two sides when $D_2^k(x)$ attains the maximum value. That is, there exist I and J such that $x_i = i$ when $1 \le i \le I$ and $x_{n_k-j+1} = n - j + 1$ when $1 \le j \le J$ ($I + J = n_k$). Otherwise, there exist x_i and two positions l_1, l_2 not in $\{x_1, \ldots, x_n\}$ such that $l_1 < x_1 < l_2$. Since one of $|l_1 - \mu_{-i}|$ and $|l_2 - \mu_{-i}|$ will be larger than $|x_i - \mu_{-i}|$, $D_2(x)$ increases if swapping x_i to position from l_1 and l_2 that are farther away from the center. (μ_{-i} is the average position of $\{x_1, \ldots x_n\} \setminus \{x_i\}$).

Then we show that $I = \frac{n_k}{2}$. Otherwise, without the loss of generality, we assume $I > \frac{n_k}{2}$. Then $\mu_k < \frac{n+1}{2}$. Therefore, $D_2(x)$ increases if swapping x_I to $n - J$, which leads to contradiction.

Theorem 7.4 *For a fixed positive integer $0 < n_k \le n$, the maximum value $D_{2,max}^{k,n}$ among all possible distributions $1 \le x_1 < x_2 < \cdots < x_{n_k} \le n$ of amino acid k can be determined using the following formulas:*

(1) n_k even, then $\mu_k = \dfrac{n+1}{2}$

$$D_{2,max}^{k,n} = \frac{1}{4}n + \frac{n_k^2 - 1}{12n} - \frac{1}{4}n_k. \tag{7.22}$$

(2) n_k odd, then $\mu_k = \dfrac{nn_k - n + 2n_k}{2n_k}$

$$D_{2,max}^{k,n} = \frac{(n_k^2 - 1)n}{4n_k^2} + \frac{n_k^2 - 1}{12n} - \frac{n_k^2 - 1}{4n_k}. \tag{7.23}$$

Proof

(1) If n_k is even, in view of Theorem 7.3, we have

$$\mu_k = \frac{n + 1}{2} \tag{7.24}$$

$$n_k n D_{2,max}^{k,n} = \sum_{i=1}^{\frac{n_k}{2}} (x_i - \frac{n + 1}{2})^2 + \sum_{i=\frac{n_k}{2}+1}^{n_k} (x_i - \frac{n + 1}{2})^2$$

$$= \sum_{i=1}^{\frac{n_k}{2}} (i - \frac{n + 1}{2})^2 + \sum_{i=1}^{\frac{n_k}{2}} (n + i - \frac{n_k}{2} - \frac{n + 1}{2})^2$$

$$= \sum_{i=1}^{\frac{n_k}{2}} (i - \frac{n + 1}{2})^2 + \sum_{i=1}^{\frac{n_k}{2}} (i + \frac{n - n_k - 1}{2})^2$$

$$= \frac{1}{6} \cdot \frac{n_k}{2}(\frac{n_k}{2} + 1)(n_k + 1) - (n + 1)\frac{\frac{n_k}{2}(\frac{n_k}{2} + 1)}{2} + \sum_{i=1}^{\frac{n_k}{2}} \frac{(n + 1)^2}{4}$$

$$+ \frac{1}{6} \cdot \frac{n_k}{2}(\frac{n_k}{2} + 1)(n_k + 1) + (n - n_k - 1)\frac{\frac{n_k}{2}(\frac{n_k}{2} + 1)}{2}$$

$$+ \sum_{i=1}^{\frac{n_k}{2}} \frac{(n - n_k - 1)^2}{4}$$

$$= \frac{n_k^3 - n_k + 3n_k(n^2 - nn_k)}{12}$$

$$\tag{7.25}$$

$$D_{2,max}^{k,n} = \frac{1}{n_k n} \frac{n_k^3 - n_k + 3n_k(n^2 - nn_k)}{12} = \frac{1}{4}n + \frac{n_k^2 - 1}{12n} - \frac{1}{4}n_k. \tag{7.26}$$

(2) If n_k is odd, then in view of Theorem 7.3

$$\mu_k = \frac{1}{n_k}\left[\sum_{i=1}^{\frac{n_k+1}{2}} i + \sum_{i=1}^{\frac{n_k-1}{2}}(n+i-\frac{n_k-1}{2})\right]$$

$$= \frac{1}{n_k}\left[2\sum_{i=1}^{\frac{n_k-1}{2}} i + \frac{n_k+1}{2} + \frac{n_k-1}{2}(n-\frac{n_k-1}{2})\right] \tag{7.27}$$

$$= \frac{nn_k - n + 2n_k}{2n_k}$$

$$n_k n D_{2,max}^{k,n} = \sum_{i=1}^{\frac{n_k+1}{2}}(i - \frac{nn_k - n + 2n_k}{2n_k})^2$$

$$+ \sum_{i=1}^{\frac{n_k-1}{2}}(n+i - \frac{n_k-1}{2} - \frac{nn_k - n + 2n_k}{2n_k})^2$$

$$= \sum_{i=1}^{\frac{n_k+1}{2}} i^2 - \frac{(n_k-1)n + 2n_k}{n_k}\sum_{i=1}^{\frac{n_k+1}{2}} i + \sum_{i=1}^{\frac{n_k+1}{2}}\frac{((n_k-1)n+2n_k)^2}{4n_k^2}$$

$$+ \sum_{i=1}^{\frac{n_k-1}{2}} i^2$$

$$+ 2(n - \frac{n_k-1}{2} - \frac{(n_k-1)n + 2n_k}{2n_k})\sum_{i=1}^{\frac{n_k-1}{2}} i$$

$$+ \sum_{i=1}^{\frac{n_k-1}{2}}(n - \frac{n_k-1}{2} - \frac{(n_k-1)n + 2n_k}{2n_k})^2$$

$$= \frac{(n_k^2-1)(n_k^2 - 3nn_k + 3n^2)}{12n_k}$$

$$\tag{7.28}$$

$$D_{2,max}^{k,n} = \frac{(n_k^2-1)(n_k^2 - 3nn_k + 3n^2)}{12nn_k^2} = \frac{(n_k^2-1)n}{4n_k^2} + \frac{n_k^2-1}{12n} - \frac{n_k^2-1}{4n_k}.$$

$$\tag{7.29}$$

Corollary 7.1 *Let c be the maximum length of the sequences in the dataset. Suppose n_k, the number of occurrences of the amino acid k, is a fixed even integer. In this case, the maximum value $D^k_{2,max}$ in the dataset can be calculated using the following expression:*

$$D^k_{2,max} = \frac{1}{4}(c - n_k) + \frac{n_k^2 - 1}{12c}. \tag{7.30}$$

Proof Let $f(n) = D^{k,n}_{2,max} = \frac{1}{4}n + \frac{n_k^2 - 1}{12n} - \frac{1}{4}n_k$ for $n_k \leq n \leq c$. Our goal is to find the maximum value of $f(n)$. It can be shown that $f(n)$ attains its minimum value at $n = \sqrt{\frac{n_k^2 - 1}{3}}$. For values of n greater than $\sqrt{\frac{n_k^2 - 1}{3}}$, $f(n)$ increases as n increases. Since $\sqrt{\frac{n_k^2 - 1}{3}} < n_k \leq n$, the maximum value of $f(n)$ occurs at $n = c$. Therefore,

$$D^k_{2,max} = \frac{1}{4}(c - n_k) + \frac{n_k^2 - 1}{12c}. \tag{7.31}$$

Corollary 7.2 *Let c represent the maximum length of the sequences in the dataset. Let n_k denote the fixed odd integer representing the number of occurrences of the amino acid k. Then the value of $D^k_{2,max}$ in the dataset can be determined as follows:*

$$D^k_{2,max} = \frac{(n_k - 1)(n_k + 1)(c - n_k)}{4n_k^2} + \frac{n_k^2 - 1}{12c}. \tag{7.32}$$

Proof Let us define the function $f(n) = D^{k,n}_{2,max} = \frac{(n_k^2 - 1)n}{4n_k^2} + \frac{n_k^2 - 1}{12n} - \frac{n_k^2 - 1}{4n_k}$ for $n_k \leq n \leq c$. In this case, the minimum value of $f(n)$ occurs at $n = \frac{n_k}{\sqrt{3}}$. As n increases beyond $\frac{n_k}{\sqrt{3}}$, $f(n)$ increases. Since $\frac{n_k}{\sqrt{3}} < n_k \leq n$, $f(n)$ reaches its maximum value at $n = c$. Therefore,

$$D^k_{2,max} = \frac{(n_k - 1)(n_k + 1)(c - n_k)}{4n_k^2} + \frac{n_k^2 - 1}{12c}. \tag{7.33}$$

The last step involves calculating the minimum value $D^k_{2,min}$ of the second normalized moment. This value is obtained when the amino acid k occupies the initial n_k positions.

Corollary 7.3 *Let $0 < n_k \le n$ be fixed positive integers. Consider a distribution of the amino acid k where $x_i = i$ for $1 \le i \le n_k$. In this distribution, the values of x_i represent the positions of the amino acid k within the sequence. Then:*

(1)

$$D_2^{k,n}(x_1, \ldots, x_{n_k}) = \frac{n_k^2 - 1}{12n}. \tag{7.34}$$

(2) *$D_2^{k,n}(x_1, \ldots, x_{n_k}) \le D_2^{k,n}(y_1, \ldots, y_n)$ where $1 \le y_1 < \cdots < y_{n_k} \le n$ is any distribution of the amino acid k.*

Proof We only need to prove (1) since (2) is obvious.

$$
\begin{aligned}
D_2^{k,n}(x_1, \ldots, x_n) &= \frac{1}{n_k n} \sum_{i=1}^{n_k} (i - 1 - \frac{n_k - 1}{2})^2 \\
&= \frac{1}{n_k n} \sum_{i=1}^{n_k} (i - \frac{n_k + 1}{2})^2 \\
&= \frac{1}{n_k n} \frac{n_k(n_k^2 - 1)}{12} = \frac{n_k^2 - 1}{12n}.
\end{aligned}
\tag{7.35}
$$

Corollary 7.4 *Let c represent the maximum length of the sequences in the dataset. Let n_k denote the fixed number representing the number of occurrences of the amino acid k. Then*

$$D_{2,min}^{k} = \frac{n_k^2 - 1}{12c}. \tag{7.36}$$

Chapter 8
New Features or Metric on Sequence Comparison

8.1 The K-mer Natural Vector Method and Its Application

In Chap. 6, we presented the natural vector approach, which is alignment-free and establishes a direct mapping between genetic sequences and vectors in a finite-dimensional space, ensuring a one-to-one correspondence. And we have shown two ways to develop the traditional natural method. The first way is to increase the order of the moments, and the second way is to take the covariance into account. In this section, we will introduce the k-mer natural vectors that contain more information and are more suitable for calculating the relationship between biological sequences [118].

The k-mer natural vector method builds upon the k-mer model, initially utilized by Blaisdell [119] for comparing genome sequences. A k-mer is a string containing k characters. A sequence can be transformed into a k-mer sequence. For example, the DNA sequence $ACGGT$ can be seen as a 2-mer sequence of length 4 $(AC)(CG)(GG)(GT)$. More generally, the sequence $s_1 s_2 \ldots s_n$ can be regarded as a sequence comprised of $n - k + 1$ k-mers $(s_1 \ldots s_k)(s_2 \ldots s_{k+1}) \ldots (s_{n-k+1} \ldots s_n)$. Kantorovitz et al. employ the frequencies of k-mers appearing in the sequences to conduct sequence comparison [120]. The k-mer model is a fast alignment-free method for the sequence comparison. Nevertheless, the limitation of the k-mer model lies in its disregard for the relationships among the k-mers within a sequence to a certain extent [121, 122]. Therefore, the k-mer natural vector method is proposed to make up for this shortcoming.

The definition of the k-mer natural vector is similar to that of the traditional natural vector. After transforming a sequence into a k-mer sequence, we can count the appearance and calculate the average position and the moments for each k-mer. We use $l_1, \ldots, l_L$ to represent all possible k-mers ($L = 4^k$ for DNA sequences and $L = 20^k$ for protein sequences). Define $s[l_j][i]$ be distance from the origin to the ith k-mer l_j in the sequence and let n_{l_j} be the number of the k-mer l_j. Then we define

© The Author(s), under exclusive license to Springer Nature Switzerland AG 2023
S. S.-T. Yau et al., *Mathematical Principles in Bioinformatics*, Interdisciplinary
Applied Mathematics 58, https://doi.org/10.1007/978-3-031-48295-3_8

$$\mu_{l_j} := \frac{\sum\limits_{i=1}^{n_{l_j}} s[l_j][i]}{n_{l_j}},$$

(8.1)

$$D_m^{l_j} := \sum_{i=1}^{n_{l_j}} \frac{(s[l_j][i] - \mu_{l_j})^m}{n_{l_j}^{m-1}(n-k+1)^{m-1}},$$

if $n_{l_j} \neq 0$ where n is the length of the original sequence ($n - k + 1 = \sum\limits_{j=1}^{L} n_{l_j}$). If $n_{l_j} = 0$, we define $\mu_{l_j} = D_1^{l_j} = \ldots = D_m^{l_j} = \ldots = 0$ in particular. A k-mer natural vector of order M can be defined as

$$(n_{l_1}, \ldots, n_{l_L}, \mu_{l_1}, \ldots, \mu_{l_L}, D_2^{l_1}, \ldots, D_2^{l_L} \ldots, D_M^{l_1}, \ldots, D_M^{l_L}).$$

(8.2)

When $k = 1$, the k-mer natural vector is identical to the original natural vector, indicating that the k-mer natural vector method is an extension of the original natural vector model.

In the case where the distribution of each k-mer varies, it is not possible for two genetic sequences to be similar solely based on containing the same set of k-mer and having the same total distance measurement. While individual subsets of numerical parameters may not be adequate for annotating genetic sequences, the collective numerical parameters are capable of characterizing each genetic sequence effectively. It can be mathematically demonstrated that there exists a one-to-one correspondence between a genetic sequence and its corresponding k-mer natural vector for a given k when the order is sufficiently high.

When conducting sequence comparisons, M is often chosen as 2 since high moments are relatively small and hardly make any contribution. Then the dimension of a k-mer natural vector is 3×4^k for DNA sequences and 3×20^k for protein sequences.

Considering that the parameter k significantly impacts the outcomes of sequence comparisons and evolutionary analyses, it becomes crucial to select an appropriate k value for different datasets. For the k-mer model, there are many works that guide the choice of k. For instance, Wu et al. [123] suggested an ideal word size for measuring dissimilarity, which is contingent upon the length of the sequences under consideration. Specifically, as the sequence length increases, it is recommended to increase the value of the optimal k, denoted by k^*. Sims et al. conducted a study [124, 125] and found that the ideal length of the k-mer falls within a range approximately bounded by $\log_4 n$, where n represents the sequence length. Additionally, they determined that the upper limit is defined by the requirement that the phylogenetic tree topology for length k should align with that of $k + 1$.

In order to determine the optimal value k^* for the k-mer natural vector method, we conducted experiments on several real datasets [78, 126]. The selection of the optimal k^* within the range of considered k values was based on the following strategy: If the phylogenetic tree obtained using value k showed a relatively stable result compared to that of $k + 1$, we chose $k^* = k$; otherwise, k^* was set to the maximum value within the range of considered k values. This approach allowed us to identify the most suitable k value for the k-mer natural vector model. Based on our inference, the optimal value k^* for the k-mer natural vector method falls within the range $[\mathrm{ceil}(\log_4 \min(L)), \mathrm{ceil}(\log_4 \max(L)) + 1]$, where L represents the set of lengths of genetic sequences considered in the phylogenetic analysis, and $\mathrm{ceil}(x)$ denotes the smallest integer greater than or equal to x. This explicit range for selecting the optimum value k^* is considerably shorter compared to the ranges considered in previous k-mer model methods. Moreover, the optimal k^* obtained using the k-mer natural vector method is smaller than those selected by other k-mer model methods [126, 127] for the same candidate dataset (18S rRNA dataset). This suggests that the k-mer natural vector method requires less computational time and can more easily extract hidden features from genetic sequences.

We can establish a distance metric to quantify the evolutionary relationships among genetic sequences because each sequence can be distinctly represented by a natural vector derived from its k-mer. There are many types of distances that can be chosen. For example, we can use the Euclidean distance to measure the dissimilarities between genetic sequences, or we can use angles to represent their similarities [127–129]. To be more specific, let v_1 and v_2 be the k-mer natural vectors of genetic sequences s_1 and s_2, respectively, and we can define the following two distances (actually d_2 is not a distance mathematically):

$$d_1(s_1, s_2) = ||v_1 - v_2||_2, \tag{8.3}$$

$$d_2(s_1, s_2) = 1 - \cos(v_1, v_2) = 1 - \frac{v_1 \cdot v_2}{||v_1||_2 ||v_2||_2}, \tag{8.4}$$

where $||.||_2$ represents Euclidean norm.

In [118], d_2 is used to construct the distance matrix, and the neighbor-joining method is applied to obtain an evolutionary tree of 40 tetrapod 18S rRNA sequences (see Fig. 8.1). k is chosen to be 6 in this problem. The phylogenetic tree depicted here exhibits four distinct clades: Birds (green), Crocodilians (blue), Mammals (red), and Amphibians (purple), with each clade accurately grouping the corresponding species together. These findings align closely with results obtained through sequence alignment and are consistent with certain phylogenetic analyses [126, 130]. (The tree plotted by ClustalW is in Fig. 8.2.)

There is another strategy to apply the k-mer natural vector. Instead of choosing a specific k, we can consider the Euclidean distance (d_1) of k-mer natural vectors with different k and sum them up with a given weight. Let $d^{(k)}$ denote the distance of k-mer natural vectors, and we can use $D(a_i, K) = \sum\limits_{i=1}^{K} a_i d^{(i)}$ to measure the dissimilarities between sequences.

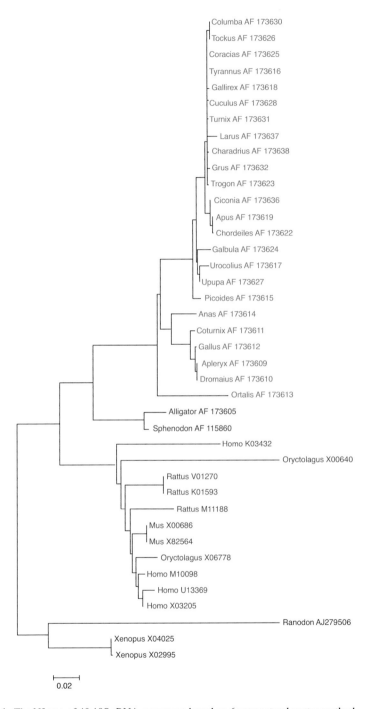

Fig. 8.1 The NJ tree of 40 18S rRNA sequences based on 6-mer natural vector method

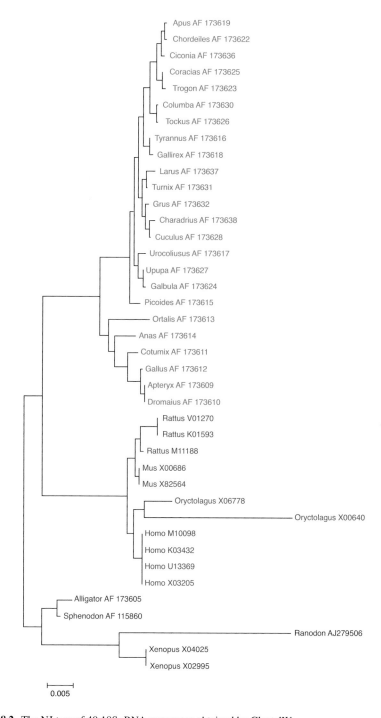

Fig. 8.2 The NJ tree of 40 18S rRNA sequences obtained by ClustalW

Table 8.1 The accuracy of $D(a_i, K)$ with different weights and K

K \ Weight	$\frac{1}{2^i}$	$\frac{1}{i^2}$
1	79.90%	79.90%
2	82.80%	82.80%
3	83.30%	83.30%
4	83.30%	83.30%
5	84.10%	84.40%
6	85.80%	86.30%
7	86.90%	87.70%
8	87.40%	88.00%
9	88.30%	85.60%

In [108], the viral classification is based on this strategy. The dataset considered consists of 7382 viral reference sequences in NCBI up to March 2020. The nearest neighborhood method (1-NN) with the leave-one-out strategy is used to check the effectiveness of the method. In the case of a viral sequence, if its closest sequence in terms of similarity belongs to the same family as itself, the classification outcome is considered accurate. The accuracy of classification can be calculated by dividing the number of correctly classified sequences by the total number of sequences. The accuracies of $D(a_i, K)$ with different K and different a_i are listed in Table 8.1. It shows that combining different k-mer is helpful for increasing the accuracy and the weight $\frac{1}{2^i}$, and $K = 9$ is a good choice for the summation.

Finally, we introduce the concept of k-mer dictionary, which may reduce the dimension of k-mer natural vectors for some cases [131]. We know that there are 4^k possible k-mers for genome sequences and 20^k possible k-mers for protein sequences, while the possible k-mers in a real dataset may be less than this number. Therefore, we can compute a k-mer dictionary that includes possible k-mers in a real dataset, which can greatly reduce the computational time. For example, titin is currently the largest known protein, and its human variant (GenBank No.: NP_001243779) consists of 34,350 amino acids [132]. Take $K = 10$, and then titin has $34,350 - 10 + 1 = 34,341$ k-mers. Therefore, even dealing with a dataset containing 1000 proteins as big as the titin, the amount of all k-mers appearing in the dataset is bounded by 4×10^7, which is far less than $20^{10} = 1.024 \times 10^{13}$. In fact, since many k-mers are duplicate, the real k-mer dictionary will be far less than the calculated bound. In [131], the size of the k-mer dictionary of a real protein dataset containing 290 proteins and a simulated dataset that has the same lengths as the real sequences in the above real dataset is calculated and listed in Table 8.2. The result further shows that there are some patterns in the real protein sequences so that the repetition of k-mers in real datasets is more common than in random sequences.

Table 8.2 The cardinalities of K-string dictionary of real and simulated dataset

K value	Cardinality (real dataset)	Cardinality (simulated dataset)
1	20	20
2	400	400
3	7186	8000
4	41703	83601
5	61792	115394
6	65733	117083
7	67214	116892
8	68182	116604
9	68898	116314
10	69450	116024
11	69895	115734
12	70255	115444
13	70551	115154
14	70804	114864
15	71012	114574
16	71188	114284
17	71343	113994
18	71482	113704
19	71607	113414
20	71720	113124

8.2 New Features Based on the Singular Value Decomposition

In this section, we will present two applications of the singular value decomposition (SVD). The first application is to transform one-hot encoding matrices to feature vectors, and the second application is to reduce the noise of the vector representation.

8.2.1 The K-mer Sparse Matrix Model and Its Applications

Yu and Huang [133] proposed the sparse matrix representation for protein primary sequences. This method can be generalized by the k-mer model [134].

We take the genetic sequence as an example. Each sequence of length L can be transformed into a matrix of size $4 \times L$ by one-hot encoding. That is, after the nucleotides are assigned to 1, 2, 3, 4, respectively, we can define $M_{ij} = 1$ if i is the assigned number of the jth nucleotide in the sequence and $M_{ij} = 0$ otherwise. Similarly, by regarding a sequence of length L as a k-mer sequence of length $L - k + 1$, we can transform each sequence into a matrix M of size $4^k \times (L - k + 1)$.

It is simple to observe that each column of matrix M contains only one element, indicating its sparsity. We can also readily verify that this transformation is bijective, preserving all the information embedded within a genetic sequence. Various powerful tools can be employed to extract characteristic information from the sparse matrix representation of k-mer. Singular value decomposition is among the available tools, and the singular value vector, composed of singular values, can be utilized to numerically represent the characteristics of a genetic sequence.

The k-mer sparse matrix M can be decomposed using singular value decomposition (SVD) as follows:

$$M = U \sum V^T, \tag{8.5}$$

where U is a real square matrix satisfying $U^T U = I_{4^k}$ (I_{4^k} is a unit matrix of rank 4^k), $\sum$ is a $4^k \times (L - k + 1)$ diagonal matrix with nonnegative singular values $\sigma_1, \sigma_2, \ldots, \sigma_{k'}$, where $k' = \min(4^k, L - k + 1)$ on the diagonal, and V^T (the transpose of V) is a real square matrix having $V^T V = I_{L-k+1}$.

Based on the given definition, we observe that the number of singular values depends on the value of k and the length L of the considered sequence. Genetic sequences often have varying lengths, and this leads to non-uniqueness in the dimension of the singular value vector, particularly when 4^k is greater than $L - k + 1$. To enable the comparison of genetic sequences using a distance measure, it is desirable to obtain a consistent dimensional singular value vector for sequences of different lengths when k is fixed. To achieve this, a slight modification is made to the k-mer sparse matrix when 4^k exceeds $L - k + 1$. Specifically, the k-mer sparse matrix is extended into a square matrix of rank 4^k by appending a zero matrix of dimensions $4^k \times (4^k - (L - k + 1))$ to the end of the k-mer sparse matrix. This modification ensures that we always obtain 4^k singular values through the SVD of k-mer sparse matrices for genetic sequences of varying lengths when k is given. This approach effectively preserves all the information encoded in the genetic sequences while enabling the characterization of genetic sequences of different lengths using singular value vectors of the same dimension.

Therefore, for each given value of k, we can construct and employ the k-mer singular value vector to capture the genetic sequence's characteristics numerically. Within this vector, the 4^k singular values are arranged in the lexicographic order of the corresponding k-mer.

In Sect. 8.1, we have shown the process of selecting the optimum value k^*. Based on this approach, we can deduce that the optimal value of k, denoted as k^*, for our k-mer sparse matrix model is approximately equal to the floor value of floor($\log_4$mean(L)), where L represents the length set of the considered genetic sequences.

We take the 18S rRNA dataset again to show the effectiveness of the method. The distance is chosen as the formula 8.4 and $k = 5$. The phylogenetic tree is shown in Fig. 8.3. The result is also satisfying.

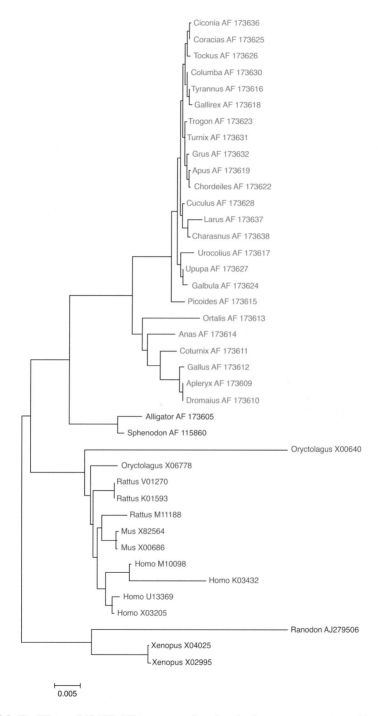

Fig. 8.3 The NJ tree of 40 18S rRNA sequences based on the 5-mer sparse matrix model

8.2.2   Noise Reduction Based on the Singular Value Decomposition

By the natural vector method or other sequence embedding techniques, we can transform a sequence into a vector and therefore transform sequences into a matrix. Let the number of sequences to be n and let the dimension of the vector to be c, and then the matrix M is c by n:

$$M = \begin{bmatrix} f_{11} & f_{12} & \cdots & f_{1n} \\ f_{21} & f_{22} & \cdots & f_{2n} \\ \cdots & \cdots & \cdots & \cdots \\ f_{c1} & f_{c2} & \cdots & f_{cn} \end{bmatrix}. \tag{8.6}$$

We can use the singular value decomposition to improve the vector representation [131, 135]. M is decomposed into three separate matrices U, $\sum$, and V using SVD, that is,

$$M = U \sum V^T, \tag{8.7}$$

where U represents the orthogonal matrix of size $c \times c$, where its columns consist of the left singular vectors of matrix M. Similarly, V represents the orthogonal matrix of size $n \times n$, with its columns comprising the right singular vectors of matrix M. The diagonal matrix $\sum$ of size $c \times n$ contains the singular values $\sigma_1 > \sigma_2 > \cdots > \sigma_{\min(c,n)}$ of matrix M arranged in descending order along its diagonal. The rank r of matrix M is determined by the number of non-zero singular values it possesses. Then the Frobenius norm of M is defined as

$$||M||_F = \sqrt{\sum_{j=1}^{r} \sigma_j^2}. \tag{8.8}$$

The Eckart–Young theorem [136] states that the distance between M and its rank-m approximations ($m \leq r$) is minimized by the approximation M_m. In this equation, M_m is expressed as the product of three matrices: U_m, $\sum_m$, and V_m^T. Here, U_m is a matrix of size $c \times m$ whose columns consist of the first m columns of matrix U. Similarly, V_m is a matrix of size $n \times m$ whose columns comprise the first m columns of matrix V. Lastly, $\sum_m$ is a diagonal matrix of size $m \times m$ whose diagonal elements correspond to the m largest singular values of matrix M. The theorem further shows how the norm of that distance is related to singular values of M:

$$||M - M_m||_F = \min_{\mathrm{rank}(X) \leq m} ||M - X|| = \sqrt{\sigma_{m+1}^2 + \cdots + \sigma_r^2}. \tag{8.9}$$

This low-rank matrix approximation can improve the quality of vector representation by discarding a substantial fraction of the noise [135]. If $\sigma_1, \ldots, \sigma_r$ are the positive singular values of M, then by using Frobenius norm, the singular vector associated with any particular singular value (i.e., σ_j) accounts for the fraction

$$\sqrt{\frac{\sigma_j^2}{\sigma_1^2 + \sigma_2^2 + \cdots + \sigma_r^2}}$$ of the data. So, choosing the m largest values ($m < r$) explains

the fraction $\sqrt{\frac{\sigma_1^2 + \sigma_2^2 + \cdots + \sigma_m^2}{\sigma_1^2 + \sigma_2^2 + \cdots + \sigma_r^2}}$ of the data, and it also allows the approximation of the

matrix from the first m singular triplets: $M_m = U_m \sum_m V_m^T$.

Determining the number m of ranked singular values that best serve to separate signal from noise within the dataset is challenging [137]. A natural idea is to fix a

number $p \in (0, 1)$, and let m to be the smallest integer satisfying $\sqrt{\frac{\sigma_1^2 + \sigma_2^2 + \cdots + \sigma_m^2}{\sigma_1^2 + \sigma_2^2 + \cdots + \sigma_r^2}} \geq$

p. It means that we make the proportion of changes in the initial matrix less than $1 - p$.

8.3 DFA7: A Novel Approach for Discriminating Intron-Containing and Intronless Genes

In this section, we present DFA7, a novel approach for categorizing genes according to their intron presence [138]. We explore three novel parameters derived from the cross-correlations among the distributions of nucleic base distances in gene sequences. By combining these new parameters with Zhang et al.'s original three parameters [139] and considering the total standard deviation, we can greatly enhance the accuracy of gene classification based on their intron status. All 7 parameters extracted from the gene can be denoted by α, β, γ, λ, θ, ϕ, and σ.

Detrended fluctuation analysis (DFA), initially proposed by Peng et al. [141], is a technique for assessing long-range power law correlations in noisy signals. This method employs a scaling analysis approach to estimate the relevant correlation parameters. In the following content, we will show how this idea contributes to feature extraction.

In a DNA sequence, the cumulative distance D_j^m represents the sum of distances between all nucleotides of the nucleic base j ($j = A, C, G, T$) and the first nucleotide (considered as the origin) within m steps. We define t_i^j as the distance from the first nucleotide to the ith nucleotide if the ith nucleotide is of type j. If the ith nucleotide is not of type j, then t_i^j is set to 0. Consequently, we can express D_j^m as the summation of t_i^j from $i = 1$ to m: $D_j^m = \sum_{i=1}^{m} t_i^j$. For example, (AGCCTCGACT) is a DNA sequence. For nucleic base C, $t_1^c = 0, t_2^c = 0, t_3^c = 2,$ $t_4^c = 3, t_5^c = 0, t_6^c = 5, t_7^c = 0, t_8^c = 0, t_9^c = 8, t_{10}^c = 0$, so $D_C^{10} = 2 + 3 + 5 + 8 = 18$. Similarly, we get $D_A^{10} = 7$, $D_G^{10} = 1 + 6 = 7$, and $D_T^{10} = 4 + 9 = 13$. Therefore, we can define three types of cumulative distances as follows:

$$\begin{cases} D_n = (D_A^n + D_G^n) - (D_C^n + D_T^n) \\ E_n = (D_A^n + D_C^n) - (D_G^n + D_T^n), \\ H_n = (D_A^n + D_T^n) - (D_C^n + D_G^n) \end{cases} \qquad (8.10)$$

where $n = 1, 2, \ldots, N$ and N represents the length of the DNA sequence, and the cumulative distances D_n, E_n, and H_n provide insights into the cross-correlation of the "position" of each nucleic base within the DNA sequence.

Next, we will build a 3×3 cumulative distance matrix using the values of D_n, E_n, and H_n. This matrix will be utilized to calculate three feature parameters: λ, θ, and ϕ. The following algorithmic steps outline the process of determining these new parameters λ, θ, and ϕ:

(1) Create a window with a width of l, where l is equal to 2^n and n can take values of $1, 2, 3, 4,$ or 5. Move this window starting from the position l_0.
(2) Compute the variation of each distribution at both ends of the window,

$$\begin{cases} \Delta D_l = D_{l_0+l} - D_{l_0} \\ \Delta E_l = E_{l_0+l} - E_{l_0} . \\ \Delta H_l = H_{l_0+l} - H_{l_0} \end{cases} \qquad (8.11)$$

(3) Iteratively shift the window from the starting position $l_0 = 1$ to $l_0 = 2$, and so on, up to $l_0 = N - l$, where N represents the length of the sequence. For each value of l_0 ranging from 1 to $N - l$, calculate the respective variations ΔD_l, ΔE_l, and ΔH_l.
(4) Define the fluctuation functions

$$\begin{cases} \rho_{DD}(l) = |(\overline{\Delta D_l \Delta D_l}) - (\overline{\Delta D_l})(\overline{\Delta D_l})| \\ \rho_{EE}(l) = |(\overline{\Delta E_l \Delta E_l}) - (\overline{\Delta E_l})(\overline{\Delta E_l})| \\ \rho_{HH}(l) = |(\overline{\Delta H_l \Delta H_l}) - (\overline{\Delta H_l})(\overline{\Delta H_l})| \end{cases} \qquad (8.12)$$

$$\begin{cases} \rho_{DE}(l) = \rho_{ED}(l) = |(\overline{\Delta D_l \Delta E_l}) - (\overline{\Delta D_l})(\overline{\Delta E_l})| \\ \rho_{DH}(l) = \rho_{HD}(l) = |(\overline{\Delta D_l \Delta H_l}) - (\overline{\Delta D_l})(\overline{\Delta H_l})| . \\ \rho_{EH}(l) = \rho_{HE}(l) = |(\overline{\Delta E_l \Delta H_l}) - (\overline{\Delta E_l})(\overline{\Delta H_l})| \end{cases} \qquad (8.13)$$

The bars represent an average over all positions l_0 in the sequence. Subsequently, the matrix of fluctuation functions can be defined as follows:

$$F = \begin{pmatrix} \rho_{DD}(l) & \rho_{DE}(l) & \rho_{DH}(l) \\ \rho_{ED}(l) & \rho_{EE}(l) & \rho_{EH}(l) \\ \rho_{HD}(l) & \rho_{HE}(l) & \rho_{HH}(l) \end{pmatrix} . \qquad (8.14)$$

Indeed, F is a real and symmetric matrix. It can be regarded as a modified version of the covariance matrix for the variables $(\Delta D_l, \Delta E_l, \Delta H_l)$. Let us denote the three eigenvalues of F as ε_1, ε_2, and ε_3, with the condition that

$\varepsilon_1 \geq \varepsilon_2 \geq \varepsilon_3$. (Readers that are not familiar with eigenvalues can refer to [140].) Through fluctuation analysis, we can derive the following result:

$$\varepsilon_1 \propto l^\lambda, \quad \varepsilon_2 \propto l^\theta, \quad \varepsilon_3 \propto l^\phi. \tag{8.15}$$

The parameters λ, θ, and ϕ are determined by the slopes observed in the log–log plots. In simpler terms, ε_i ($i = 1, 2, 3$) can be seen as a proportional function of l^j ($j = \lambda, \theta, \phi$), meaning that ε_j can be expressed as $c \times l^j$, where c is a non-zero constant. Due to the nonlinear scaling of the axes, a function in the form of $y = a \times x^b$ will appear as a straight line on a log–log graph, with b representing the slope of the line. As a result, the parameters λ, θ, and ϕ can be calculated by estimating the slope of the log–log graph corresponding to $\varepsilon_1, \varepsilon_2$, and ε_3 using the available numerical data.

(5) Estimate the slopes λ, θ, and ϕ of each log–log graph corresponding to $\varepsilon_1, \varepsilon_2$, and ε_3 computed in step (4).

Hence, by following the aforementioned five algorithmic steps, we can compute three parameters λ, θ, and ϕ for any given DNA sequence. In step (1), to enhance computational efficiency and minimize error in determining the slope ϕ, we utilize values of $l = 2^n$ ($n = 1, 2, 3, 4, 5$). These values result in a perfectly fitted line. Even in cases where linearity is not ideal, a least-squares fit of the data enables us to obtain a unique straight line while minimizing the squared error associated with the slope and intercept parameters.

In [139], three exponents α, β, and γ are extracted with the same step from the Z-curve, which is a curve representation defined as follows:

Let us consider a DNA sequence consisting of N bases. We define the number of steps as n ($n = 1, 2, \ldots, N$). To analyze the sequence, we count the cumulative occurrences of the bases A, C, G, and T in the subsequence from the first to the nth base. These cumulative counts are represented by A_n, C_n, G_n, and T_n, respectively. The Z-curve is a three-dimensional curve that comprises a series of nodes P_n ($n = 1, 2, \ldots, N$), with coordinates denoted as x_n, y_n, and z_n. It is shown that

$$\begin{cases} x_n = 2(A_n + G_n) - n \\ y_n = 2(A_n + C_n) - n \, . \\ z_n = 2(A_n + T_n) - n \end{cases} \tag{8.16}$$

The Z-curve of the DNA sequence is defined by connecting the nodes $P_0 = 0$, $P_1, \ldots, P_N$ in a sequential manner using straight lines, where $n = 1, 2, \ldots, N$, and the initial cumulative counts are $A_0 = C_0 = G_0 = T_0 = 0$.

We can extract three exponents α, β, and γ from (x_n, y_n, z_n) by the same way by which we extract λ, θ, and ϕ from (D_l, E_l, H_l). Therefore, we can combine these exponents and get a 6-dimensional feature vector. In conclusion, the inclusion of the parameter σ, representing the sample standard deviation of the other 6 features, is essential. The cumulative distances (D_l, E_l, H_l), derived directly from the original DNA sequence, capture additional sequence information that the Z-curve method lacks. By incorporating these new parameters, we can improve the accuracy of gene classification.

Following the preceding procedures, we obtain a 7-dimensional feature vector for each gene sequence. Subsequently, a machine learning technique employing a support vector machine (SVM) with a Gaussian radial basis kernel function (RBF) is employed to predict intronless and intron-containing genes solely based on the primary sequences [138].

In SVM, there are two types of parameters: the penalty parameter C and the kernel type K. The penalty parameter C acts as a regularization parameter, controlling the balance between maximizing the margin and minimizing the training error. The kernel type K is another crucial parameter. For this problem, we utilize the radial basis kernel function $K(x_i, x_j) = \exp(-\gamma||x_i - x_j||^2)$. In this chapter, the parameter γ in the radial basis function plays a significant role in determining the generalization ability of SVM by regulating the kernel function's amplitude. Therefore, optimizing the two parameters C and γ is necessary. To perform parameter optimization, a grid search approach within a limited range can be employed.

The dataset used to test this method consists of 1000 randomly selected intronless genes from UniProtKB/Swiss-Prot (release 15.1) and 1000 randomly selected intron-containing genes from Genbank database (release 170). These genes are derived from various eukaryotes such as humans, thale cress, and Mus musculus to ensure diversity and avoid similarity. Four methods, namely GENSCAN, N-SCAN, Z-curve method, and DFA7, are compared in terms of classification results (refer to Table 8.3 and Fig. 8.4). For the classification process, a five-fold cross-validation strategy is employed, and the parameter values (C, γ) are selected as $(16, 2^{-6})$ using a grid search approach. The results indicate that DFA7 exhibits the best performance among the four methods.

8.4   The Lempel–Ziv Complexity and Its Application in Sequence Comparison

In the following section, we will present the notion of Lempel–Ziv complexity, a valuable metric for quantifying the uncertainty of a sequence [142]. We will then utilize this measure to establish the distance between two genomes [143].

Table 8.3 Accuracy of different methods on 2000 mixed prokaryotic and eukaryotic genes (%)

Methods	1	2	3	4	5	Average
GENSCAN	76.50	74.00	76.75	78.25	77.50	76.60 ± 1.61
N-SCAN	82.50	81.75	83.75	80.25	81.50	81.95 ± 1.29
Z-Curve	88.75	87.25	85.25	83.75	85.75	86.15 ± 1.90
DFA7	94.75	93.50	92.75	91.75	92.50	93.05 ± 1.14

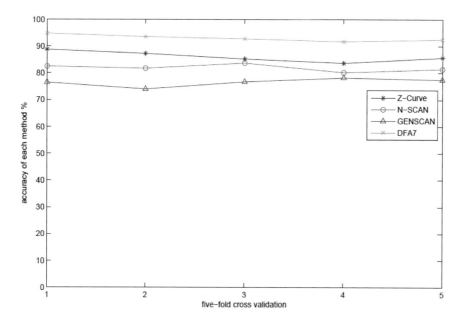

Fig. 8.4 The accuracy comparison of DFA7 and other three methods on 2000 mixed prokaryotic and eukaryotic genes

Let S be a sequence defined over an alphabet Ω, $L(S)$ be the length of S, $S(i)$ denotes the ith element of S, and $S(i, j)$ defines the substring of S composed of the elements of S between positions i and j (inclusive). For DNA case, $\Omega = \{A, C, G, T\}$, if $S = AACGTCGTCG$, then $L(S) = 10$, $S(4) = G$, and $S(4, 7) = GTCG$.

The Lempel–Ziv complexity of a sequence S can be quantified as the minimum number of steps needed to synthesize it within a specific process

$$H(S) = S(1 : i_1) \cdot S(i_1+1 : i_2) \cdot \cdots \cdot S(i_{k-1}+1 : i_k) \cdot \cdots \cdot S(i_{m-1}+1 : N). \quad (8.17)$$

At each step, two operations are allowed: copying the longest fragment from the part of S that has already been synthesized, or generating a new symbol that ensures the uniqueness of each component $S(i_{k-1} + 1 : i_k)$.

More specifically, at each step k, the sequence S is extended by concatenating a fragment $S(i_{k-1} + 1 : i_k)$. The length of this fragment is 1 if some symbol at position $i_{k-1} + 1$ occurs for the very first time. Otherwise, this fragment is obtained by copying from the prefix $S(1 : i_{k-1})$ and adding an additional symbol. The Lempel–Ziv complexity is the number of concatenating components in this process. For example, given a DNA sequence $S = AACGTACCATTG$, the Lempel–Ziv schema of synthesis gives the following components: $H(S) = A \cdot \; < A > C \cdot G \cdot T \cdot \; < AC > C \cdot \; < A > T \cdot \; < T > G$ (here <> means that the copied part from the prefix and $\cdot$ separate each component), and the corresponding complexity $C_{LZ}(S) = 7$.

Another example is that, given a DNA sequence $R = CTAGGGGACTTAT$, the Lempel–Ziv schema of synthesis gives the following components: $H(R) = C \cdot T \cdot A \cdot G \cdot < GGG > A \cdot < CT > T \cdot < A > T$, and $C_{LZ}(R) = 7$. Note that, during one concatenating component, the part from the prefix can be continually copied many times, like G here.

Ziv and Lempel [144] called the complexity decomposition of a sequence S following the above schema the exhaustive history of S and mathematically proved that every sequence S has a unique exhaustive history.

The Lempel–Ziv complexity provides a powerful tool for measuring the similarity between two DNA sequences [145]. Given two sequences S and R, consider the sequence SR and its Lempel–Ziv complexity. By definition, the number of components needed to build R when appended to S is $C_{LZ}(SR) - C_{LZ}(S)$. This number will be less than or equal to $C_{LZ}(R)$ because at any given step of the production process of R (in building the sequence SR) we use a larger search space due to the existence of S. Therefore, if R is more similar to S than T, then we would expect $C_{LZ}(SR) - C_{LZ}(S)$ to be smaller than $C_{LZ}(ST) - C_{LZ}(S)$. Here we adopt a similarity measure between two sequences P and Q as

$$d(P, Q) = \frac{C_{LZ}(PQ) - C_{LZ}(P) + C_{LZ}(QP) - C_{LZ}(Q)}{\frac{1}{2}(C_{LZ}(PQ) + C_{LZ}(QP))}. \tag{8.18}$$

Due to its successful application in the phylogenetic analysis of complete mammalian mitochondrial genomes, the Lempel–Ziv complexity has gained significant usage and recognition [145].

The application of the Lempel–Ziv complexity in measuring the distance between multi-segmented genomes is extended in reference [143]. By defining the distance between two sequences and employing the Hausdorff distance, it becomes possible to generalize this measurement to multi-segmented genomes. Additionally, the method can be further extended using a modified Hausdorff distance, as described in [146]. The modified Hausdorff distance is defined as $d_2(A, B) = \frac{1}{n} \sum_{a \in A} d(a, B)$, and then the modified Hausdorff distance is defined by

$$MHD(A, B) = \max\{d_2(A, B), d_2(B, A)\}. \tag{8.19}$$

8.5 An Information-Based Network Approach for Protein Classification

In the present section, we introduce a novel approach for protein classification, which utilizes an information-based network methodology [147]. This innovative method is rooted in the principles of information theory and network analysis. Initially, it involves the construction of a discrete map representing the amino acid sequences. The subsequent mathematical intricacies of this novel approach are delineated as follows.

In our analysis, we examine the amino acid sequence data of proteins. A protein sequence consists of 20 distinct types of amino acids, which can be regarded as a discrete time series with 20 possible states. To facilitate further analysis, each amino acid is mapped uniquely to an integer value $b \in \{1, 2, \ldots, 20\}$. By employing this mapping scheme, the entire protein sequence is transformed into a discrete time series representation.

For a pair of integer time series X and Y, where X has a length of M and Y has a length of N ($M \leq N$), we define Y_i ($1 \leq i \leq N - M + 1$) as the ith segment of length M in Y. The mutual information rate [148, 149] between X and Y_i ($1 \leq i \leq N - M + 1$) is calculated as

$$I(X; Y_i) = \sum_{\alpha \in S_X, \beta \in S_{Y_i}} p(x = \alpha, y = \beta) \log \frac{p(x = \alpha, y = \beta)}{p(x = \alpha) p(y = \beta)}. \tag{8.20}$$

In the equation, S_X and S_{Y_i} represent the state sets (subsets of integers from 1 to 20) of X and Y_i, respectively. It is important to note that the summation in this equation is position-free, as it encompasses all possible combinations of states between X and Y_i. The mutual information rate is a probability expectation of logarithmic ratios, and it solely relies on the probability distribution of the states involved, rather than the specific values of the states themselves.

By varying i from 1 to $N - M + 1$, we obtain a series of mutual information rates: $I_1, \ldots, I_{N-M+1}$. To determine the maximum mutual information rate between sequences X and Y, we select the largest value from this series:

$$I_{XY,\max} = \max_{1 \leq i \leq N-M+1} I(X; Y_i). \tag{8.21}$$

To calculate the maximum mutual information rates for each pair of integer sequences, we generate a matrix that represents these rates. In this matrix, denoted as I_{XY}, each element corresponds to the maximum mutual information rate between the sequences X and Y.

The mutual information rate provides a measure of the mutual relationships between sequences. By using the maximum mutual information matrix as the adjacency matrix, we can construct a protein network. In this network, each protein is represented as a node, and the connections between nodes are determined by their mutual relationships. The protein network is considered a weighted network, where the weights are given by the values in the adjacency matrix. It is important to note that the protein network is undirected because the mutual information rate is symmetric, meaning that the mutual information rate between protein X and protein Y is the same as the mutual information rate between protein Y and protein X [148].

$$a_{XY} = a_{YX} = I_{XY,\max} = I_{YX,\max}. \tag{8.22}$$

The maximum mutual information rate $I_{XY,\max}$ between two arbitrary sequences X (with a length of M) and Y (with a length of N) in the dataset is attained when a specific value of k exists, where $1 \le k \le N - M + 1$:

$$I_{XY,\max} = I(X; Y_k). \tag{8.23}$$

Due to the upper bound imposed by the entropies in information theory [148, 149], the mutual information rate is subject to the following relationship:

$$I(A; B) \le \min\{H(A), H(B)\}, \tag{8.24}$$

where $H(A) = -\sum_{\alpha \in S_A} p(a = \alpha)\log p(a = \alpha)$, $H(B) = -\sum_{\beta \in S_B} p(b = \beta)\log p(b = \beta)$, and A and B denote two general sequences of the same length. Therefore, we have

$$I_{XY,\max} = I(X; Y_k) \le \min\{H(X), H(Y_k)\}. \tag{8.25}$$

In addition, Y_k is a subsequence of Y, and we therefore have [148]

$$H(Y_k) \le H(Y), \tag{8.26}$$

and the inequality holds:

$$I_{XY,\max} \le \min\{H(X), H(Y)\}. \tag{8.27}$$

In order to enable a universal comparison between nodes, the adjacency matrix is normalized by dividing each element by the maximum entropy. Assuming there are N proteins in the database, the adjacency element between the ith and jth proteins is transformed as follows:

$$a_{ij} = \frac{I_{ij,\max}}{\max_{1 \le q \le N} H_q}, \tag{8.28}$$

where H_q denotes the entropy of the qth protein. By dividing each element of the adjacency matrix by the maximum entropy, the resulting values are bounded between 0 and 1, ensuring a standardized range for the elements.

For the normalized adjacency matrix, a mutation threshold is defined to filter the matrix. This threshold is determined by multiplying a constant with the maximum adjacency element. In other words, the mutative threshold can be expressed as

$$T_c = c \cdot \max_{i,j} a_{ij}. \tag{8.29}$$

The multiplicity, denoted as c, ranges from 0.1 to 1. For each value of c, denoted as T_c, it is used as a threshold to filter the adjacency matrix. Elements below this threshold are set to zero, while the rest remain unchanged. By examining the adjacency elements of the filtered matrix, the connected components of the network can be identified. In matrix form, the filtered adjacency matrix is raised to powers with exponents ranging from 1 to $N - 1$, where N represents the number of nodes in the network, which is also the size of the adjacency matrix. In a network, the connected components refer to the largest sets of nodes where all members are interconnected through at least one path [150]. To obtain these connected components, the adjacency matrix raised to the power s is summed from $s = 1$ to $s = N - 1$.

$$A_{\text{sum}} = \sum_{i=1}^{N-1} A^i. \tag{8.30}$$

The connected components are defined as the largest sets of nodes where all elements in the sum matrix A_{sum} are positive. Adding new nodes to these components would result in the violation of this property. Intuitively, through reversible transformations, the sum matrix A_{sum} can be transformed into a block diagonal form, where each block represents a connected component.

$$\begin{pmatrix} D_1 & 0 & \cdots \\ 0 & D_2 & \cdots \\ \vdots & \vdots & \ddots \end{pmatrix}. \tag{8.31}$$

Each connected component of the network is represented by a positive diagonal matrix, denoted as D_i ($i = 1, 2, \dots$).

By varying the multiplicity c, the connected components of the network undergo changes. The connected components obtained with a higher threshold exhibit stronger connections among their nodes compared to those obtained with a lower threshold. As a result, the connected components obtained with a higher threshold are encompassed by the connected components obtained with a lower threshold. Based on the inclusion or exclusion of these connected components, we create protein classifications. The presence or absence of specific connected components within a given threshold range determines the classification of proteins.

We apply this method to a dataset consisting of 35 NADH dehydrogenase proteins encoded by mitochondrial genes from 35 diverse mammalian species [47].

The classification results of the 35 NADH dehydrogenase proteins are depicted in Fig. 8.5, where the proteins are represented by their corresponding mammalian species. The figure showcases the classification of NADH dehydrogenase proteins based on their respective orders, namely Carnivora, Artiodactyla, Perissodactyla, Lagomorpha, Rodentia, Proboscidea, Primate, and Eulipotyphla. Within these

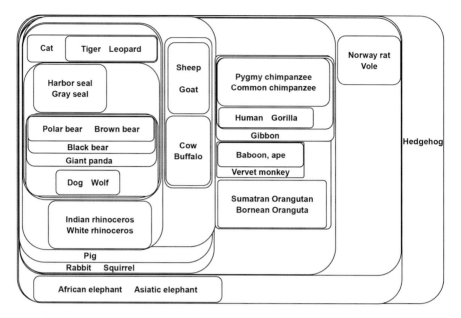

Fig. 8.5 The figure illustrates the component graph of mitochondrial proteins from 35 mammalian species. It showcases the connected components derived from the adjacency matrix, which has been filtered using various thresholds ($T_c = c \cdot A_{\max}$, $c \in [0, 1]$). Each set in the figure represents a connected component, and it demonstrates that components obtained with higher thresholds are inclusive of those obtained with lower thresholds. The proteins are represented according to their respective mammalian species

orders, four primary clusters can be observed: Primate, Rodentia (including Norway rat and Vole), Eulipotyphla (including Hedgehog), and other non-primate orders.

Within the primates, the classification of NADH dehydrogenase proteins reveals three distinct groups. The first group belongs to the Homininae family, which includes the Pygmy chimpanzee, Common chimpanzee, Gorilla, and Human ($c = 0.85$). The second group consists of the Cercopithecidae family (baboon, Vervet monkey) and the Hominidae family (ape). Lastly, the third group belongs to the Hominidae family and specifically includes the Ponginae subgroup, represented by the Sumatran orangutan and Bornean orangutan ($c = 0.8$).

The classification of non-primate orders is also accurate. The Carnivora order is divided into distinct groups: Phocidae (Gray seal and Harbor seal, $c = 0.98$), Ursidae (Brown bear, Polar bear, Black bear, and Giant panda, $c = 0.94$), Canidae (Dog and Wolf, $c = 1$), and Felidae (Cat and Tiger, Leopard, $c = 0.94$). The proteins within the Perissodactyla order (Indian rhinoceros and White rhinoceros, $c = 0.93$) are also properly grouped. Additionally, the proteins of the Artiodactyla order are classified into the Bovinae subfamily (Cow and Buffalo, $c = 0.97$), the Caprinae subfamily (Sheep and Goat, $c = 0.96$), and the Suidae family (Pig). It is worth noting that the rabbit, belonging to the Lagomorpha order, is a distinct species.

Within the diverse range of non-primate orders, the Carnivora order serves as the central group, surrounded by the Perissodactyla order and the Artiodactyla order.

The clustering results obtained from this method exhibit consistency with other approaches, such as libSVM, which relies on machine learning techniques [151]. Furthermore, this method not only uncovers the mutual relationships between proteins but also identifies clusters based on their connections, offering additional insights compared to machine learning methods.

References

1. M.S. Waterman, Introduction to computational biology: maps, sequences and genomes, Chapman & Hall/CRC, New York, 1995.
2. R. Tyagi, Computational molecular biology, Discovery Publishing House Pvt. Ltd., Delhi, 2009.
3. C.J. Benham, Sites of predicted stress-induced DNA duplex destabilization occur preferentially at regulatory loci, Proceedings of the National Academy of Science USA, 90: 2999–3003, 1993.
4. T.E. Creighton, W.H. Freeman, et al., Protein folding, New York, 1992.
5. D. Haig, L. Hurst, A quantitative measure of error minimization in the genetic code, Journal of Molecular Evolution, 33: 412–417, 1991.
6. S.J. Freeland, L.D. Hurst, The genetic code is one in a million, Journal of Molecular Evolution, 47: 238–248, 1998.
7. J. Konecny, M. Schöniger, G.L. Hofacker, Complementary coding conforms to the primeval comma-less code, Journal of Theoretical Biology, 173: 263–270, 1995.
8. M. Eigen, B.F. Lindemann, M. Tietze, R. Winkler-Oswatitsch, A. Press, A. von Haeseler, How old is the genetic code? Statistical geometry of tRNA provides an answer, Science, 244: 673–679, 1989.
9. C. Kanz, P. Aldebert, et al., The embl nucleotide sequence database, Nucleic Acids Research, 33: D29-D33, 2005.
10. H. Sugawara, T. Abe, T. Gojobori, Y. Tateno, Ddbj working on evaluation and classification of bacterial genes in insdc, Nucleic Acids Research, 35: D13-D15, 2007.
11. D.A. Benson, M. Cavanaugh, K. Clark, I. Karsch-Mizrachi, J. Ostell, K.D. Pruitt, E.W. Sayers, GenBank, Nucleic Acids Research, 46: D41-D47, 2018.
12. P. Rodriguez-Tome, P. Stoehr, G. Cameron, T. Flores, The European Bioinformatics Institute (EBI) databases, Nucleic acids research, 24, 6–12, 1996.
13. E.C. Kamau, G. Winter(Eds.), Common Pools of Genetic Resources: Equity and Innovation in International Biodiversity Law, Routledge, 2013.
14. The UniProt Consortium, Reorganizing the protein space at the universal protein resource (UniProt), Nucleic Acids Research, 40: D71-D75, 2012.
15. C. H. Wu, L. S. Yeh, H. Huang, et al., The Protein Information Resource, Nucleic acids research, 31: 345–347, 2003.
16. K. Najarian, S. Najarian, S. Gharibzadeh, C.N. Eichelberger, Systems Biology and Bioinformatics: A Computational Approach, CRC Press, 2009.

© The Author(s), under exclusive license to Springer Nature Switzerland AG 2023
S. S.-T. Yau et al., *Mathematical Principles in Bioinformatics*, Interdisciplinary Applied Mathematics 58, https://doi.org/10.1007/978-3-031-48295-3

17. R.D. Finn, J. Mistry, J. Tate, P. Coggill, A. Heger, J.E. Pollington, O.L. Gavin, P. Gunesekaran, G. Ceric, K. Forslund, L. Holm, E.L. Sonnhammer, S.R. Eddy, A. Bateman, The pfam protein families database, Nucleic Acids Research, 38: D211-D222, 2010.

18. C.J.A. Sigrist, E. Castro, L. Cerutti, B.A. Cuche, N. Hulo, A. Bridge, L. Bougueleret, I. Xenarios, New and continuing developments at PROSITE, Nucleic Acids Research, 41: D344-D347, 2013.

19. F.C. Bernstein, T.F. Koetzle, G.J. Williams, E.E. Meyer, M.D Brice, J.R. Rodgers, O. Kennard, T. Shimanouchi, M. Tasumi, The protein data bank: a computer-based archival file for macromolecular structures, Journal of Molecular Biology, 112: 535–542, 1977.

20. A.G. Murzin, S.E. Brenner, T. Hubbard, C. Chothia, Scop: a structural classification of proteins database for the investigation of sequences and structures, Journal of Molecular Biology, 247: 536–540, 1995.

21. F.M. Pearl, C.F. Bennett, J.E. Bray, A.P. Harrison, N. Martin, A. Shepherd, I. Sillitoe, J. Thornton, C.A. Orengo, The CATH database: an extended protein family resource for structural and functional genomics, Nucleic acids research, 31, 452–455, 2003.

22. C.A. Orengo, A.D. Michie, S. Jones, D.T. Jones, M.B. Swindells, J.M. Thornton, Cath-a hierarchic classification of protein domain structures, Structure, 5: 1093–1108, 1997.

23. L. Holm, P. Rosenstrom, Dali server: conservation mapping in 3D, Nucleic Acids Research, 38: W545-W549, 2010.

24. T. Madej, K.J. Addess KJ, J.H. Fong, et al. MMDB: 3D structures and macromolecular interactions. Nucleic acids research, 40(Database issue): D461-D464., 2012.

25. K.C. Wong, Computational Biology and Bioinformatics: Gene Regulation, CRC Press, 2016.

26. A. Isaev, Introduction to mathematical methods in bioinformatics, Springer, 2005.

27. M. O. Dayhoff, R. M. Schwartz, B. C. Orcutt, A model of Evolutionary Change in Proteins, Atlas of protein sequence and structure, 5: 345–358. 1978.

28. S. Henikoff, J. Henikoff, Amino acid substitution matrices from protein blocks, Proceedings of the National Academy of Sciences of the United States of America, 89: 10915–10919, 1992.

29. S. B. Needleman and C. D. Wunsch, A general method applicable to the search for similarities in the amino acid sequence of two proteins, Journal of molecular biology, 48: 443–453, 1970.

30. T. F. Smith and M. S. Waterman, Identification of common molecular subsequences, Journal of molecular biology, 147: 195–197, 1981.

31. D. J. Lipman, W. R. Pearson, Rapid and sensitive protein similarity searches, Science. 227: 1435–1441, 1985.

32. S. Altschul, W. Gish, W. Miller, E. Myers, D. J. Lipman, Basic local alignment search tool, Journal of Molecular Biology, 215: 403–410, 1990.

33. D. G. Higgins, P. M. Sharp, CLUSTAL: a package for performing multiple sequence alignment on a microcomputer, Gene, 73: 237–244, 1988.

34. R.H. Strunk, D.S. Stoer, Time series analysis and its applications: with R examples, Springer, third edition, 2010.

35. C. Burge, S. Karlin, Prediction of complete gene structures in human genomic DNA, Journal of Molecular Biology, 268: 78–94, 1997.

36. M.Q. Zhang, Identification of protein coding regions in the human genome by quadratic discriminant analysis, Proceedings of the National Academy of Science USA, 94: 565–568, 1997.

37. S.A. Marhon, S.C. Kremer, Gene prediction based on DNA spectral analysis: a literature review, Journal of Computational Biology, 18: 1–28, 2011.

38. S. Tiwari, S. Ramachandran, A. Bhattacharya, S. Bhattachrya, R. Ramaswamy, Prediction of probable genes by Fourier analysis of genomic sequences, CABIOS, 113: 263–270, 1997.

39. D. Anastassiou, Frequency-domain analysis of biomolecular sequences, Bioinformatics, 16: 1073–1081, 2000.

40. C. Yin, S.S.-T. Yau, A Fourier characteristic of coding sequences: origins and a non-Fourier approximation, Journal of Computational Biology, 9: 1153–1165, 2005.

41. K. Shah, A. Krishnamachari, On the origin of three base periodicity in genomes, Biosystems, 107: 142–144, 2012.

42. C. Yin, S.S.-T. Yau, Prediction of protein coding regions by the 3-base periodicity analysis of a DNA sequence, Journal of Theoretical Biology, 247: 687–694, 2007.

43. C. Yin, D. Yoo, S.S.-T. Yau, Tracking the 3-Base periodicity of protein-coding regions by the nonlinear tracking-differentiator, Proceedings of the 45th IEEE Conference on Decision & Control, 2094–2097, 2006.

44. C. Yin, D. Yoo, S.S.-T. Yau, Denoising the 3-Base periodicity walks of DNA sequences in gene finding, Journal of Medical and Bioengineering, 2: 80–83, 2013.

45. C. Yin, S.S.-T. Yau, Numerical representation of DNA sequences based on genetic code context and its applications in periodicity analysis of genomes, IEEE Symposium on Computational Intelligence in Bioinformatics & Computational Biology, 2: 223–227, 2008.

46. X. Jiang, S.S.-T. Yau, A novel analysis model for DNA sequences, International Conference on BioMedical Engineering and Informatics, 1: 24–28, 2008.

47. B. Zhao, R. He, S.S.-T. Yau, A new distribution vector and its application in genome clustering, Molecular Phylogenetics and Evolution, 59: 438–443, 2011.

48. S.S.-T. Yau, J. Wang, A. Niknejad, C. Lu, N. Jin, Y. Ho, DNA sequence representation without degeneracy, Nucleic Acids Research, 31: 3078–3080, 2003.

49. C. Yin, S.S.-T. Yau, An improved model for whole genome phylogenetic analysis by Fourier transform, Journal of Theoretical Biology, 382: 99–110, 2015.

50. T. Hoang, C. Yin, S.S.-T. Yau, Numerical encoding of DNA sequences by chaos game representation with application in similarity comparison, Genomics, 108: 134–142, 2016.

51. T. Hoang, C. Yin, H. Zheng, C. Yu, R. He, S.S.-T. Yau, A new method to cluster DNA sequences using Fourier power spectrum, Journal of Theoretical Biology, 372: 135–145, 2015.

52. F. Sievers, D.G. Higgins, Clustal Omega for making accurate alignments of many protein sequences, Protein Sci, 27: 135–145, 2018.

53. R. Dong, Z. Zhu, C. Yin, R. He, S.S.-T. Yau, A new method to cluster genomes based on cumulative Fourier power spectrum, Gene, 673: 239–250, 2018.

54. E. Hamori, J. Ruskin, H curves, a novel method of representation of nucleotide series especially suited for long DNA sequences, Journal of Biological Chemistry, 258: 1318–1327, 1983.

55. M.A. Gates, Simpler DNA sequence representations, Nature, 316: 219, 1985.

56. L. Liu, Y. Ho, S.S.-T. Yau, Clustering DNA sequences by feature vectors, Molecular Phylogenetics and Evolution, 41: 64–69, 2006.

57. C. Yu, Q. Liang, C. Yin, R. He, S.S.-T. Yau, A novel construction of genome space with biological geometry, DNA Research, 17: 155–168, 2010.

58. C. Yu, M. Deng, S.S.-T. Yau, DNA sequence comparison by a novel probabilistic method, Information Sciences, 181: 1484–1492, 2011.

59. C.M. Cover, J.A. Thomas, Elements of information theory, John Wiley and Sons, NY, 1991.

60. R.R. Sokal and C.D. Michener, A statistical method for evaluating systematic relationships, University of Kansas science bulletin, 38: 1409–1438, 1958.

61. S.S.-T. Yau, C. Yu, R. He, A protein map and its application, DNA and Cell Biology, 27: 241–250, 2008.

62. J. Fauchere, V. Pliska, Hydrophobic parameters of amino-acid side-chains from the partitioning of N-acetyl-amino acid amides, European Journal of Medicinal Chemistry, 18: 369–375, 1983.

63. C. Yu, S.Y. Cheng, R. He, S.S.-T. Yau, Protein map: An alignment-free sequence comparison method based on various properties of amino acids, Gene, 486: 110–118, 2011.

64. X. Xia, W.H. Li, What amino acid properties affect protein evolution? Journal of Molecular Evolution, 47: 557–564, 1998.

65. P.H.A. Sneath, Relations between chemical structure and biological activity, Journal of Theoretical Biology, 12: 157–195, 1966.

66. K. Tian, X. Yang, Q. Kong, C. Yin, R. He, S.S.-T. Yau, Two dimensional Yau-Hausdorff distance with applications on comparison of DNA and protein sequences, PLoS ONE, 10: e0136577, 2015.

67. D.P. Huttenlocher, G.A. Klanderman, W.J. Rucklidge, Comparing images using the Hausdorff distance, IEEE Transactions on Pattern Analysis and Machine Intelligence, 15: 850–863, 1993.

68. D.P. Huttenlocher, K. Kedem, J.M. Kleinberg, On dynamic Voronoi diagrams and the minimum Hausdorff distance for point sets under Euclidean motion in the plane, Proceedings of the eighth annual symposium on Computational geometry, 110–119, 1992.

69. L.P. Chew, M.T. Goodrich, D.P. Huttenlocher, K. Kedem, J.M. Kleinberg, D. Kravets, Geometric pattern matching under Euclidean motion, Computational Geometry, 7: 113–124, 1997.

70. G. Rote, Computing the minimum Hausdorff distance between two point sets on a line under translation, Information Processing Letters, 38: 123–127, 1991.

71. B. Li, Y. Shen, B. Li, A new algorithm for computing the minimum Hausdorff distance between two point sets on a line under translation, Information Processing Letters, 106: 52–58, 2008.

72. P.D. Hebert, A. Cywinska, S.L. Ball, J.R. deWaard, Biological identifications through DNA barcodes, Proc. Biol. Sci., 270: 313–321, 2003.

73. Jeffrey, H. Joel, Chaos game representation of gene structure, Nucleic Acids Research, 18: 2163–2170, 1990.

74. T. Hoang, C. Yin, S.S.-T. Yau, Splice sites detection using chaos game representation and neural network, Genomics, 112: 1847–1852, 2020.

75. A. Fiser, G. E. Tusnády, I. Simon, Chaos game representation of protein structures, Journal of Molecular Graphics, 12: 302–304, 1994.

76. Z. Yu, V. Anh, K. Lau, Chaos game representation of protein sequences based on the detailed HP model and their multifractal and correlation analyses, Journal of Theoretical Biology, 226: 341–348, 2004.

77. Z. Sun, S. Pei, R. He, S.S.-T. Yau, A novel numerical representation for proteins: Three-dimensional Chaos Game Representation and its Extended Natural Vector, Computational and Structural Biotechnology Journal, 18: 1904–1913, 2020.

78. M. Deng, C. Yu, Q. Liang, R. He, S.S.-T. Yau, A novel method of characterizing genetic sequences: genome space with biological distance and applications, PLoS ONE, 6: e17293, 2011.

79. H. Musto, S. Caccio, H. Rodriguez, G. Bernardi, Compositional constraints in the extremely GC-poor genome of Plasmodium falciparum, Mem Inst Oswaldo Cruz, 92: 835–841, 1997.

80. D.G. Mead, Newton's Identities, The American Mathematical Monthly, 99: 749–751, 1992.

81. C. Yu, T. Hernandez, H. Zheng, S.C. Yau, H. Huang, R. He, J. Yang, S.S.-T. Yau, Real time classification of viruses in 12 dimensions, PLoS ONE, 8: e64328, 2013.

82. D. Baltimore, Expression of animal virus genomes, Bacteriological reviews, 35: 235–241, 1971.

83. D. Baltimore, Viral genetic systems, Transactions of the New York Academy of Sciences, 33: 327–332, 1971.

84. D. Baltimore, The strategy of RNA viruses, Harvey lectures, 70: 57–74, 1974.

85. H. Guo, W.S. Mason, C.E. Aldrich, J.R. Saputelli, D.S. Miller, A.R. Jilbert, J.E. Newbold, Identification and characterization of Avihepadnaviruses isolated from exotic Anseriformes maintained in captivity, Journal of Virology, 79: 2729–2742, 2005.

86. X. Chen, X. Ruan, Q. Zhao, H. Li, Molecular identification of badnavirus in Dracaena sanderiana from Hubei of China, Scientia Agricultura Sinica, 42: 2002–2009, 2009.

87. H. Huang, C. Yu, H. Zheng, T. Hernandez, S.C. Yau, R. He, J. Yang, S.S.-T. Yau, Global comparison of multiple-segmented viruses in 12-dimensional genome space, Molecular Phylogenetics and Evolution, 81: 29–36, 2014.

88. D. Liu, W. Shi, Y. Shi, D. Wang, H. Xiao, W. Li, Y. Bi, Y. Wu, X. Li, J. Yan, W. Liu, G. Zhao, W. Yang, Y. Wang, J. Ma, Y. Shu, F. Lei, G.F. Gao, Origin and diversity of novel avian

influenza A H7N9 viruses causing human infection: phylogenetic, structural, and coalescent analyses, Lancet, 381: 1926–1932, 2013.

89. M. Barrett, M.J. Donoghue, E. Sober, Against consensus, Systematic Zoology, 40: 486–493, 1991.

90. H. Zheng, C. Yin, T. Hoang, R. He, J. Yang, S.S.-T. Yau, Ebolavirus classification based on natural vectors, DNA and Cell Biology, 34: 418–428, 2015.

91. M. Levitt, Nature of the protein universe, Proceedings of the National Academy of Science USA, 106: 11079–11084, 2009.

92. E.V. Koonin, Y.I. Wolf, G.P. Karev, The structure of the protein universe and genome evolution, Nature, 420: 218–223, 2002.

93. E.V. Koonin, Metagenomic sorcery and the expanding protein universe, Nature Biotechnology, 25: 540–542, 2007.

94. C. Yu, M. Deng, S.C. Cheng, S.C. Yau, R. He, S.S.-T. Yau, Protein space: A natural method for realizing the nature of protein universe, Journal of Theoretical Biology, 318: 197–204, 2013.

95. H. Mellor, P.J. Parker, The extended protein kinase C superfamily, Biochem J, 332: 281–292, 1998.

96. X. Zhao, X. Wan, R. He, S.S.-T. Yau, A new method for studying the evolutionary origin of the SAR11 clade marine bacteria, Molecular Phylogenetics and Evolution, 98: 271–279, 2016.

97. H. Luo, Evolutionary origin of a streamlined marine bacterioplankton lineage, ISME Journal, 9: 1423–1433, 2015.

98. J. Viklund, T. Ettema, S. Andersson, Independent genome article and phylogenetic reclassification of the oceanic SAR11 clade, Molecular Biology and Evolution, 29: 599–615, 2012.

99. J. Viklund, J. Martijn, T. Ettema, S. Andersson, Comparative and phylogenomic evidence that the alphaproteobacterium HIMB59 is not a member of the oceanic SAR11 clade, PLoS ONE, 8: e78858, 2013.

100. J. Gower, G. Ross, Minimum spanning trees and single linkage cluster analysis, J. Roy. Stat. Soc., 18: 54–64, 1969.

101. N. Saitou, M. Nei, The neighbor-joining method: a new method for reconstructing phylogenetic trees, Mol. Biol. Evol., 4: 406–425, 1987.

102. A. Enfissi, J. Codrington, J. Roosblad, M. Kazanji, D. Rousset, Zika virus genome from the Americas, Lancet, 387: 227–228, 2016.

103. Y. Li, L. He, R. He, S.S.-T. Yau, Zika and Flaviviruses phylogeny based on the alignment-Free natural vector method, DNA and Cell Biology, 36: 109–116, 2017.

104. R.S. Lanciotti, A.J. Lambert, M. Holodniy, S. Saavedra, L.C. Signor, Phylogeny of Zika virus in Western hemisphere, 2015, Emerging Infectious Diseases, 22: 933–935, 2016.

105. X. Zhao, K. Tian, R. He, S.S.-T. Yau, Establishing the phylogeny of Prochlorococcus with a new alignment-free method, Ecology and Evolution, 7: 11057–11065, 2017.

106. K. Tian, X. Zhao, S.S.-T. Yau, Convex hull analysis of evolutionary and phylogenetic relationships between biological groups, Journal of Theoretical Biology, 456: 34–40, 2018.

107. X. Zhao, K. Tian, R. He, S.S.-T. Yau, Convex hull principle for classification and phylogeny of eukaryotic proteins, Genomics, 111: 1777–1784, 2019.

108. N. Sun, S. Pei, L. He, C. Yin, R. He, S.S.-T. Yau, Geometric construction of viral genome space and its applications, Computational and Structural Biotechnology Journal, 19: 4226–4234, 2021.

109. A.M. Martinez, A.C. Kak, PCA versus LDA, IEEE Trans Pattern Anal Mach Intell, 23:228–233, 2001.

110. X. Jiao and S. Pei and Z. Sun and J. Kang and S.S.-T. Yau, Determination of the nucleotide or amino acid composition of genome or protein sequences by using natural vector method and convex hull principle, Fundamental Research, 1: 559–564, 2021.

111. R. Zhao, S. Pei, S.S.-T. Yau, New Genome Sequence Detection via Natural Vector Convex Hull Method, IEEE/ACM Transactions on Computational Biology and Bioinformatics, 19: 1782–1793, 2022.

112. S.S.-T. Yau, W. Mao, M. Benson, R. He, Distinguishing proteins from arbitrary amino acid sequences, Scientific Reports, 5: 1–8, 2015.
113. The UniProt Consortium, Activities at the universal protein resource (UniProt), Nucleic Acids Research, 42: D191-D198, 2014.
114. B. Kuhlman, G. Dantas, G.C. Ireton, G. Varani, B.L. Stoddard, D. Baker, Design of a novel globular protein fold with atomic-level accuracy, Science, 302: 1364–1368, 2008.
115. Y.L. Chan, M.S. Brown, D. Qin, N. Handa, D.K. Bishop, The third exon of the budding yeast meiotic recombination gene HOP2 is required for calcium-dependent and recombinase Dmc1-specific stimulation of homologous strand assimilation, Journal of Biological Chemistry, 289: 18076–18086, 2014.
116. D. Deng, C. Xu, P. Sun, J. Wu, C. Yan, M. Hu, N. Yan, Crystal structure of the human glucose transporter GLUT1, Nature, 510: 121–125, 2014.
117. T.W. Nilsen, B.R. Graveley, Expansion of the eukaryotic proteome by alternative splicing, Nature, 463: 457–463, 2010.
118. J. Wen, R. Chan, S.C. Yau, R. He, S.S.-T. Yau, K-mer natural vector and its application to the phylogenetic analysis of genetic sequences, Gene, 546: 25–34, 2014.
119. B.E. Blaisdell, A measure of the similarity of sets of sequences not requiring sequence alignment, Proceedings of the National Academy of Science USA, 83: 5155–5159, 1986.
120. M.R. Kantorovitz, G.E. Robinson, S. Sinha, A statistical method for alignment-free comparison of regulatory sequences, Bioinformatics, 23: 249–255, 2007.
121. X.W. Yang, T.M. Wang, A novel statistical measure for sequence comparison on the basis of k-word counts, Journal of Theoretical Biology, 318: 91–100, 2013.
122. H.J. Yu, Segmented K-mer and its application on similarity analysis of mitochondrial genome sequences, Gene, 518: 419–424, 2013.
123. T.J. Wu, Y.H. Huang, L.A. Li, Optimal word sizes for dissimilarity measures and estimation of the degree of dissimilarity between DNA sequences, Bioinformatics, 21: 4125–4132, 2005.
124. G.E. Sims, S.R. Jun, G.A. Wu, S.H. Kim, Alignment-free genome comparison with feature frequency profiles (FFP) and optimal resolutions, Proceedings of the National Academy of Science USA, 106: 2677–2682, 2009.
125. G.E. Sims, S.R. Jun, G.A. Wu, S.H. Kim, Whole-genome phylogeny of mammals: evolutionary information in genic and non-genic regions, Proceedings of the National Academy of Science USA, 106: 17077–17082, 2009.
126. R.H. Chan, T.H. Chan, H.M. Yeung, R.W. Wang, Composition vector method based on maximum entropy principle for sequence comparison, IEEE/ACM Transactions on Computational Biology and Bioinformatics, 9: 79–87, 2012.
127. J. Qi, B. Wang, B.L. Hao, Whole proteome prokaryote phylogeny without sequence alignment: a k-string comparison approach, Journal of Molecular Evolution, 58: 1–11, 2004.
128. J. Wen, Y. Zhang, A 2D graphical representation of protein sequence and its numerical characterization, Chemical Physics Letters, 476: 281–286, 2009.
129. G.W. Stuart, M.W. Berry, An SVD-based comparison of nine whole eukaryotic genomes supports a coelomate rather than ecdysozoan linkage, BMC Bioinformatics, 5: 204, 2004.
130. S.B. Hedges, K.D. Moberg, L.R. Maxson, Tetrapod phylogeny inferred from 18S and 28S ribosomal RNA sequence and a review of the evidence for amniote relationships, Molecular Biology and Evolution, 7: 607–633, 1990.
131. C. Yu, R. He, S.S.-T. Yau, Protein sequence comparison based on K-string dictionary, Gene, 529: 250–256, 2013.
132. A. Minajeva, M. Kulke, J.M. Fernandez, W.A. Linke, Unfolding of titin domains explains the viscoelastic behavior of skeletal myofibrils, Biophysical Journal, 80: 1442–1451, 2001.
133. H.J. Yu, D.S. Huang, Novel 20-D descriptors of protein sequence and its applications in similarity analysis, Chemical Physics Letters, 531: 261–266, 2012.
134. J. Wen, Y. Zhang, S.S.-T. Yau, K-mer Sparse matrix model for genetic sequence and its applications in sequence comparison, Journal of Theoretical Biology, 363: 145–150, 2014.
135. G.W. Stuart, K. Moffett, J.J. Leader, A comprehensive vertebrate phylogeny using vector representations of protein sequences from whole genomes, Molecular Biology and Evolution, 19: 554–562, 2002.

136. C. Eckart, G. Young, The approximation of one matrix by another of lower rank, Psychometrika, 1: 211–218, 1936.
137. M.W. Berry, Z. Drmac, E.R. Jessup, Matrices, vector spaces, and information retrieval, SIAM Review, 41: 335–362, 1999.
138. C. Yu, M. Deng, L. Zheng, R. He, J. Yang, S.S.-T. Yau, DFA7, a new method to distinguish between intron-containing and intronless genes, PLoS ONE, 9: e101363, 2014.
139. C.T. Zhang, Z.S. Lin, M. Yan, R. Zhang, A novel approach to distinguish between intron-containing and intronless genes based on the format of Z curves, Journal of Theoretical Biology, 192: 467–473, 1998.
140. S. Lang, Eigenvectors and Eigenvalues. In: Introduction to Linear Algebra. Undergraduate Texts in Mathematics. Springer, New York, NY, 1986.
141. C.K. Peng, S.V. Buldyrev, A.L. Coldberger, S.L. Havlin, F. Sciortino, Longrange correlations in nucleotide sequences, Nature, 356: 168–170, 1992.
142. A. Lempel, J. Ziv, On the complexity of finite sequences, IEEE Transactions on Information Theory, 22:75–81, 1976.
143. C. Yu, R. He, S.S.-T. Yau, Viral genome phylogeny based on Lempel-Ziv complexity and Hausdorff distance, Journal of Theoretical Biology, 348: 12–20, 2014.
144. J. Ziv, A. Lempel, A universal algorithm for sequential data compression, IEEE Transactions on Information Theory, 23: 337–343, 1977.
145. H.H. Otu, K. Sayood, A new sequence distance measure for phylogenetic tree construction, Bioinformatics, 19: 2122–2130, 2003.
146. M.P. Dubuisson, A.K. Jain, A modified Hausdorff distance for object matching, Proceedings of the 12th IAPR International Conference on Pattern Recognition, Conference A: Computer Vision & Image Processing, 1: 566–568, 1994.
147. X. Wan, X. Zhao, S.S.-T. Yau, An information-based network approach for protein classification, PLoS ONE, 12: e0174386, 2017.
148. Y. Zhou, The basics of information theory, 3rd Edition, Beijing University of Aeronautics and Astronautics Press, 2006.
149. K. Hlavackova, M. Palus, M. Vejmelka, J. Bhattacharya, Causality detection based on information-theoretic approached in time series analysis, Physics Reports, 441: 1–46, 2007.
150. S.Z. Raina, J.J. Faith, T.R. Disotell, H. Seligmann, C.B. Stewart, D.D. Pollock, Evolution of base-substitution gradients in primate mitochondrial genomes, Genome Research, 15: 665–673, 2005.
151. C. Chang, C. Lin, LibSVM: A Library for support vector machines, ACM Transactions on Intelligent Systems & Technology, 2: 1–27, 2011.